P9-CMW-546

FROM VITAMIN A TO ZINC,
FROM AMINO ACIDS TO WILD YAM,
FROM ACIDOPHILUS TO YOHIMBE

THE COMPLETE GUIDE TO VITAMINS, HERBS, AND SUPPLEMENTS

- An overview of more than 150 commonly available natural supplements, including vitamins, minerals, herbs, and amino acids
- How to treat everyday ailments with natural substances
- Simple, straightforward advice on how to design a personal supplement program
- What form to buy, how much to take, and possible side-effects and interactions with other medications and supplements
- Latest medical information and nutritional studies

Good health is in your hands!

ATTENTION: ORGANIZATIONS AND CORPORATIONS
Most Avon Books paperbacks are available at special quantity
discounts for bulk purchases for sales promotions, premiums, or
fund-raising. For information, please call or write:

Special Markets Department, HarperCollins Publishers, Inc.,
10 East 53rd Street, New York, N.Y. 10022–5299.
Telephone: (212) 207–7528. Fax: (212) 207-7222.

THE
COMPLETE GUIDE TO VITAMINS HERBS
AND
SUPPLEMENTS

THE HOLISTIC PATH TO GOOD HEALTH

WINIFRED CONKLING WITH
DAVID Y. WONG, M.D., CONSULTING EDITOR

A LYNN SONBERG BOOK

AVON BOOKS
An Imprint of HarperCollinsPublishers

This book contains advice and information relating to health care. It is not intended to replace medical advice and should be used to supplement rather than replace regular care by your doctor. It is recommended that you seek your physician's advice before embarking on any medical program or treatment. All efforts have been made to assure the accuracy of the information contained in this book as of the date of publication. The publisher and the authors disclaim liability for any medical outcomes that may occur as a result of applying the methods suggested in this book.

AVON BOOKS
An Imprint of HarperCollins*Publishers*
195 Broadway
New York, NY, 10007

Copyright © 2006 by Lynn Sonberg Book Associates
ISBN-13: 978-0-06-076066-3
ISBN-10: 0-06-076066-4
www.avonbooks.com

All rights reserved. No part of this book may be used or reproduced in any manner whatsoever without written permission, except in the case of brief quotations embodied in critical articles and reviews. For information address Avon Books, an Imprint of HarperCollins Publishers.

First Avon Books paperback printing: January 2006

Avon Trademark Reg. U.S. Pat. Off. and in Other Countries, Marca Registrada, Hecho en U.S.A.
HarperCollins® is a registered trademark of HarperCollins Publishers Inc.

Printed in the U.S.A.

20 19 18 17

If you purchased this book without a cover, you should be aware that this book is stolen property. It was reported as "unsold and destroyed" to the publisher, and neither the author nor the publisher has received any payment for this "stripped book."

For my father-in-law
Dan Rak

≈ CONTENTS

Chapter 3: Amino Acids and Other Nutrition Supplements

Chapter 4: Homeopathic Remedies   119

PART 2:
PRESCRIPTION FOR HEALING
155

⪡ INTRODUCTION:
Mother Nature's Miracles

If you eat a balanced diet and follow a healthy lifestyle, do you really need to take nutrition supplements? A few years ago, most doctors would have congratulated you on your clean living and told you to skip the supplements to save money. Today, however, most experts realize that almost everyone—including you—can benefit from taking some supplements.

Why the about-face? Simply put, my colleagues and I in the medical profession have a better understanding of nutrition and biochemistry than we did just a generation ago. Advances in our knowledge about nutrition and disease have underscored the essential role vitamins, minerals, herbs, and other substances play in good health. In addition, mounting evidence has shown that taking supplements may help prevent heart disease, cancer, osteoporosis, and other chronic diseases. I recommend supplements to virtually every patient I see at my Health Integration Centers in Torrance and Santa Monica, California.

Of course, supplements can't make up for an unhealthy diet and poor lifestyle choices. To enjoy optimal health, you need to eat plenty of fruits, vegetables, and whole grains, while avoiding excessive amounts of processed foods, sugar, and fat. But even once-skeptical doctors now agree, you would be well served to also include a multivitamin and selected nutrients in your daily routine.

THE IMPORTANCE OF SUPPLEMENTS

You already know that good nutrition is the foundation of good health. For the body to function properly, it must have a sufficient supply of more than forty key nutrients. While few Americans suffer from severe deficiency diseases such as scurvy or rickets, the Council for Responsible Nutrition reports that most Americans do not consume adequate amounts of many nutrients, including vitamins A, B6, B12, C, E, thiamin, riboflavin, and folic acid, and the minerals calcium, chromium, iron, magnesium, selenium, and zinc. These inadequate intakes may not trigger a disease directly caused by a nutrient deficiency, but they leave the body vulnerable to other diseases and chronic conditions, such as heart disease and cancer, among many others.

This link between nutrition and health has been an area of extensive study in the past decade. Scientific investigations into the medicinal value of vitamins, minerals, herbs, phytochemicals (chemicals in plants), enzymes, hormones, and other natural supplements have led researchers and physicians to appreciate the health-enhancing properties of these compounds.

In other words, food is not only fuel for the body, but medicine as well. Unfortunately, the typical American diet is deficient in nutrients and fiber and high in fat, cholesterol, and preservatives. Alas, even if you include lots of fresh fruits, vegetables, and whole grains in your diet, you may not be consuming as many nutrients as you think you are. Most nonorganic foods grow in nutritionally deficient soil, which generates nutritionally deficient produce. Furthermore, between the time the food leaves the field and reaches your dinner plate, additional nutrients are destroyed by food processing, storing, and cooking.

To make matters worse, your body may not be able to take advantage of all the nutrients you do consume. Emotional and physical stress can cause a breakdown of your immune system and make you susceptible to invasion by harmful microorganisms, resulting in countless problems, from the common cold to ulcers to heart disease, all of which can hinder how your body absorbs and uses nutrients. Stress also increases your body's requirements for many nutrients, especially the water-soluble vitamins such as vitamin C and the B vitamins. That's why, for example, you can buy vitamin combinations sold as "stress formulas," which contain the B vitamins and often vitamin C as well.

In addition, the process of aging increases the body's demands for nutrients. As you age, your body loses some of its ability to assimilate the nutrients you consume. For example, as you grow older, you may experience declining levels of many important nutrients, including amino acids (such as methionine and cysteine), antioxidants (such as coenzyme Q10 and vitamin E), and DHEA (a precursor for many hormones). Ironically, your body's demand for these nutrients increases at the same time your physical reserves drop. Supplementation is one solution. While it cannot stop aging, it can help make the inevitable process easier and healthier.

ALTERNATIVE MEDICINE COMES OF AGE

Alternative or natural medicine is based on one simple truth: The human body has considerable power to heal itself. Wounds heal, rashes disappear, and illnesses run their course. As a healer, I strive to support the natural healing process of the body. It is essential to view yourself as a whole—body, mind, and spirit—rather than focusing on the part that is sick or injured.

Practitioners of conventional medicine too often focus on controlling symptoms of disease or injury, using drugs and surgery to deal with specific complaints or conditions.

In recent years, natural medicine has come into its own. For more than a decade, the sales of supplements, including vitamins and herbs, have risen steadily and dramatically. According to the *Nutrition Business Journal*, Americans spent more than $20 billion on supplements in 2003. Clearly, the American public has learned to appreciate the health benefits associated with nutrition supplements.

In addition, alternative healing is finally receiving the respect it deserves from practitioners of mainstream medicine. Many medical organizations that in the past had spoken out against natural medicine now endorse many of the same recommendations that practitioners of natural medicine have been making for decades. For the last half-century, naturopaths have recommended that people eat high-fiber foods, exercise on a regular basis, reduce stress, and cut back on the intake of refined sugars, fats, and processed foods, but it is only in the last decade or so that many conventional practitioners support these same suggestions.

Likewise, nutritional and herbal medicine and other

natural methods of healing are being taught in mainstream hospital centers and medical schools, places where the techniques were dismissed not long ago. Even the renowned National Institutes of Health opened the National Center for Complementary and Alternative Medicine in 1998, and began funding research on various alternative medicine techniques.

Natural medicine is taking its place next to traditional medicine, especially in the treatment of relatively minor conditions. If you develop a serious illness or suffer a major injury, by all means head for the doctor, but to stave off a cold or deal with a chronic health problem, you might want to try an alternative treatment. No single system of medicine provides all the answers all the time. A cooperative approach to medical care, which includes both traditional treatments and natural remedies, may provide the best care for your overall health.

The Complete Book of Vitamins, Herbs, and Supplements for Health and Healing provides simple and straightforward advice on how to design a supplement program for general health, as well as how to choose supplements to address specific common medical problems. I believe vitamins, minerals, herbs, and other nutrition supplements play an essential role in establishing good health. This book will demystify the world of nutritional supplements so that you can use them to enhance your overall health.

—David Y. Wong, M.D.

☙ USING THIS BOOK

When it comes to choosing supplements, you have a wide range of choices, from vitamin A to zinc, amino acids to wild yam, acidophilus to yohimbe. It's easy to feel overwhelmed and uninformed, but this book will help you identify the supplements that will be the most beneficial for you and to design a supplement program customized to meet your individual needs.

The book is divided into two parts—Part 1: Understanding, Choosing, and Using Healing Supplements and Part 2: Prescription for Healing. Part 1 of this book provides an overview of more than 150 commonly available natural supplements, including chapters on vitamins and minerals, herbs, amino acids and other nutrition supplements, and homeopathic remedies. Each entry provides basic information about what the supplement does, what form to buy, how much to take, and special warnings or possible interactions with other medications or supplements.

People who want to jump right in and begin customizing their supplement program right away may want to turn to Chapter 5 for advice on choosing a multivitamin–mineral supplement as well as additional core nutrients for overall health. These readers can then refer to the material in the earlier chapters if they want specific information about a certain vitamin, mineral, herb, or other nutrients. Consult Chapter 6 for an explanation of recommended intake amounts—Recommended Dietary Allowances (RDAs) and Dietary Reference Intakes (DRIs).

Part 2 is an alphabetical series of entries describing the most common everyday health problems and how nutrition supplements can be used in their treatment. It includes a basic description of each condition, as well as advice on when to seek professional medical care. Since "an ounce of prevention is worth a pound of cure," each entry also includes practical tips on prevention whenever appropriate.

The information presented in this book is safe and accurate, but even the most seemingly mundane conditions demand professional medical care when complications arise. If you are seriously ill or do not respond to the treatments listed here, promptly seek the care of a trained medical professional.

PART 1:

Understanding, Choosing, and Using Healing Supplements

Understanding,
choosing, and using
Healing Supplements

⬅ CHAPTER ONE
Vitamins and Minerals

Vitamins and minerals are essential for good health, but which ones do *you* need for optimal health? This chapter will help you understand the biological importance of various vitamins and minerals, and it will provide details on how to safely use these nutritional supplements. This information can be used in conjunction with the health information described in Chapter 7, as you design a supplement plan to meet your specific health needs.

HOW MUCH IS ENOUGH—AND TOO MUCH?

If you're like most people, you probably don't eat what you should every day. You may reach for burgers and fries or cookies and cakes with some regularity, making you wonder whether you should supplement your daily bread with a daily vitamin.

In virtually all cases, the answer is yes. A well-balanced diet is a cornerstone of good health, but multivitamins and nutrition supplements can come in handy when you want to make up for dietary failings. A daily vitamin provides peace of mind that you are getting enough nutrients, even on the days when you succumb to temptation.

Faced with the possibility of nutritional shortfalls, some people may be tempted to load up with vitamin and mineral supplements. But the "if some is good, more is better" approach does not apply to vitamins. Large doses of vitamins over long periods of time can trigger side effects, some of which can be serious.

When using nutritional supplements, you will take either a daily dose, which can be taken at a given amount on an ongoing basis, or a therapeutic dose, which should be used for a limited time to give the body a boost in either preventing or managing an illness. To avoid overdose, take the higher amount only during the course of the illness or as long as recommended on the product label.

You should also be aware that vitamins can be either fat- or water-soluble. Fat-soluble vitamins are stored in the body; megadoses of these vitamins can build up in the body and cause dangerous side effects. The fat-soluble vitamins are vitamin A, vitamin D, vitamin E, and vitamin K. Water-soluble vitamins are stored in smaller amounts in the body and must be consumed more often. They include the B vitamins and vitamin C. Excessive amounts of water-soluble vitamins are excreted from the body in the urine.

The following section provides an alphabetical list of vitamins and minerals, including information on food sources of the nutrient, signs of deficiency, medical uses, dosages, side effects, and any known drug interactions. You can refer back to these entries from Chapter 7 when you want specific information on the use of these nutrients for the treatment of medical problems.

VITAMINS AND MINERALS A TO Z

Biotin

Biotin—also known as vitamin B7 and vitamin H—is a member of the B vitamin family. Its primary functions in the body are to assist with the metabolism of fats, carbohydrates, and proteins, and to help with cell growth and facilitate the utilization of the other B vitamins. Biotin also has proved helpful in lowering and controlling the blood sugar levels in people with either insulin-dependent or non-insulin-dependent diabetes.

Good Food Sources: Soy, whole grains, egg yolk, almonds, walnuts, oatmeal, mushrooms, broccoli, bananas, peanuts, liver, kidney, milk, legumes, sunflower seeds, and nutritional yeast.

Signs of Deficiency: Signs of biotin deficiency include depression, hair loss, high blood sugar, anemia, loss of appetite, insomnia, muscle cramps, nausea, and a sore tongue. In addition, low biotin levels have been linked to seborrheic dermatitis in infants; biotin's role in causing this condition in adults has not been established.

Biotin deficiency is very rare, in part because this vitamin can be manufactured by the intestines from other foods. Long-term use of antibiotics, however, can hinder production of biotin and lead to deficiency symptoms. Signs of deficiency are also seen in people who regularly consume raw egg whites, which contain a protein called avidin that prevents the absorption of biotin into the blood.

Uses of Biotin: Biotin is used in the treatment of diabetes (page 219).

Dosage Information: The adult RDA is 100 to 200 micrograms; the therapeutic dose is 200 micrograms. Purchase either a multivitamin–mineral supplement or a B-complex formula that contains biotin. Most people do not need to take a separate biotin supplement unless they are treating diabetes, in which case it is recommended you do so under a doctor's guidance.

Possible Side Effects: Biotin is a nontoxic, water-soluble vitamin; if excessive amounts are taken, it is excreted in the urine without causing adverse effects. People with diabetes who are taking insulin may need to decrease their insulin dosage if they take more than 4 milligrams of biotin daily; diabetics should be under a doctor's care.

Possible Interactions: Biotin works in conjunction with the other B vitamins. Substances that can interfere with bioavailability of biotin include antibiotics, saccharin, and sulfa drugs.

Boron

Boron is a trace mineral that plays an important role in maintaining healthy bones, cartilage, and joints. It is also essential for the absorption of calcium, magnesium, and phosphorus. In addition, boron has been credited with enhancing brain function and promoting mental alertness.

Good Food Sources: Raisins, almonds, prunes, most noncitrus fruits, and leafy green vegetables. (The level of boron in various foods depends on the level of boron in the soil.)

Signs of Deficiency: No cases of boron deficiency have been reported. Low levels of boron have been associated with an increased risk of osteoporosis in postmenopausal women.

Uses of Boron: Boron is used in the treatment of osteoarthritis (page 176) and osteoporosis (page 296).

Dosage Information: Boron is not included in many multivitamin–mineral formulas because the federal government has not established an RDA for boron. For general health, look for a multivitamin that contains 1.5 to 3 milligrams of boron. If you have osteoarthritis or osteoporosis, consider taking 3 to 9 milligrams of boron daily in tablet or powder form. Look for sodium borate or boron chelates for osteoporosis; look for sodium tetraborate decahydrate for the treatment of osteoarthritis.

Possible Side Effects: No adverse effects have been noted when boron is taken at recommended levels (at or below 9 milligrams per day).

Possible Interactions: Boron may help the body conserve its supply of calcium, magnesium, and phosphorus. It may increase estrogen levels in postmenopausal women taking supplemental estrogen; it has not been found to raise estrogen levels in postmenopausal women not taking estrogen or in men or premenopausal women.

Calcium

Calcium is perhaps best known for its critical role in the formulation of bones and teeth. While more than 99 percent of the calcium in the body is stored in the bones, the remaining 1 percent plays an essential role in other body functions, such as muscle growth, transmission of nerve impulses, blood clotting, and a regular heartbeat.

According to an August 1997 report issued by the National Academy of Sciences, most American adults and children get only about half the calcium they need from the foods they eat. If you do not consume enough calcium through your diet, calcium is stripped from the bones to continue essential body functions. Over time, this will result in weak bones and a condition known as osteoporosis.

Good Food Sources: Milk, cheese, green leafy vegetables, salmon (with bones), almonds, blackstrap molasses, brewer's yeast, broccoli, kale, kelp, sesame seeds, tofu, and yogurt.

Signs of Deficiency: Signs of calcium deficiency include muscle cramps, heart palpitations, high blood pressure, nervousness, tooth decay, rickets, numbness in the legs and arms, brittle nails, and aching joints.

Uses of Calcium: Calcium is used to treat anxiety (page 173), arteriosclerosis (page 183), cancer of the colon (page 198), heart attack and cardiovascular disease (page 255), hemorrhoids (page 262), hypertension (page 266), insomnia (page 279), menopausal symptoms (page 287), osteoporosis (page 296), Parkinson's disease (page 299), and premenstrual syndrome (page 302).

Dosage Information: The adult DRI is 800 milligrams for men, 1,200 milligrams for women, and 1,500 milligrams for pregnant and lactating women. Because calcium is more effective when the body receives it in smaller amounts, divide your daily intake into two or three doses. If possible, take calcium one hour before or two hours after meals and before bedtime, rather than in one megadose.

Calcium supplements are always combined with other chemicals or salts. Common forms include calcium carbonate, calcium citrate, calcium gluconate, or calcium lactate. The difference among these forms is the percentage of elemental calcium in the supplement and the absorbency. The higher the percentage of elemental calcium in the supplement, the fewer capsules, tablets, or chews you will need to take to reach the optimal calcium intake.

The forms of calcium that are best absorbed by the body are calcium citrate and calcium carbonate. You can test the absorption effectiveness of your calcium supplement if you place a calcium pill in a glass of warm water and shake it. Let the mixture sit for twenty-four hours. If the calcium has not dissolved after twenty-four hours, the absorbency rate is poor. Switch to another brand.

Because calcium and magnesium work closely together in the body, many experts recommend taking the two nutrients together. Combination supplements are available; some with a ratio of 2:1 (calcium to magnesium) and others a 1:1. Experts disagree as to the best ratio, although 2:1 seems to be preferred.

Avoid natural oyster shell calcium, dolomite, and bone meal products, which have a history of lead contamination. Calcium carbonate has the lowest lead content.

Possible Side Effects: Taking too much calcium can cause constipation or calcium deposits in the soft tissues. Do not take calcium supplements if you have kidney stones or kidney disease. Avoid Tums with calcium as a calcium source because the antacid neutralizes the acid needed for calcium absorption.

Possible Interactions: Drugs used to treat epilepsy and other seizure disorders can lead to calcium deficiency; if you take these medications, discuss the need for calcium supplementation with your doctor.

If you take iron supplements, take your calcium supplement at least two hours after the iron, because calcium inhibits the effectiveness of both nutrients.

For calcium to be absorbed properly, adequate vitamin D is needed. If you get twenty to thirty minutes of direct sunlight exposure per day, you do not need a vitamin D supplement. The elderly and bedbound patients often need to take a combined calcium and vitamin D supplement.

Chromium

Chromium is an essential trace mineral that helps the body maintain healthy levels of cholesterol and blood sugar, in addition to assisting with the synthesis of cholesterol, fats, and proteins. The body needs the hormone insulin to get glucose from the blood into the tissues where it can be used for energy, and chromium increases the sensitivity of tissues to the action of insulin. If you are deficient in chromium, your body will have trouble maintaining normal glucose utilization. (Chromium itself has no effect on glucose; it only works together with insulin to drive sugar from blood to tissue.)

About 90 percent of the population does not get enough chromium from food. In addition, a high-sugar diet can increase the excretion of chromium, leading to obesity and diabetes. Some experts believe widespread chromium deficiency has contributed to the surge in Type II diabetes in the United States.

Good Food Sources: Brewer's yeast, beer, brown rice, grains, cereals, liver, legumes, peas, and molasses. (Refining and processing foods dramatically reduces chromium levels in foods.)

Signs of Deficiency: Signs of deficiency include blood sugar fluctuations and high cholesterol.

Uses of Chromium: Chromium is used to treat acne (page 158), diabetes (page 219), glaucoma (page 245), obesity (page 293), and psoriasis (page 306).

Dosage Information: The recommended level for chromium is 50 to 200 micrograms daily; the therapeutic dose is 200 micrograms daily. The more carbohydrate you eat, the more chromium you need. Take chromium with food and with vitamin C to increase absorption.

The preferred forms are chelated tablets, such as chelated chromium picolinate, which is chromium chelated with the natural amino acid metabolite called picolinate. Picolinate allows chromium to enter the cells more efficiently. Another form, chromium polynicotinate (chromium chelated to niacin), is also effective. Chromium is often part of a high-quality multivitamin–mineral formula.

Possible Side Effects: No toxicity has been noted at doses of 50 to 300 micrograms per day. Some people develop a rash or feel light-headed when taking chromium. If taken regularly at levels of 1,000 micrograms or higher, kidney and liver damage is likely. Because chromium can affect blood sugar levels, people with diabetes should consult their physicians before taking chromium supplements and closely monitor their blood sugar levels.

Possible Interactions: Vitamin C helps increase the absorption of chromium. Refined sugars, white flour products, and lack of exercise deplete chromium levels.

Copper

This essential trace mineral has an essential role in the formation of bone, red blood cells, and hemoglobin, and it is necessary for the proper absorption and utilization of iron. It also plays a part in energy production, regulation of heart rate and blood pressure, fertility, taste sensitivity, skin and hair coloring, and the healing process.

Good Food Sources: Seafood, organ meats, blackstrap molasses, nuts, seeds, green vegetables, black pepper, cocoa, and water that is carried via copper pipes.

Signs of Deficiency: Signs of deficiency include brittle hair, anemia, high blood pressure, heart arrhythmias, infertility, and skeletal defects. Copper deficiency is uncommon, but it can occur in people who take a zinc supplement without increasing their copper intake, because zinc (as well as vitamin C and calcium) can interfere with copper absorption. Deficiency may also occur in people who have Crohn's disease, celiac disease, albinism, or in infants who are not breast-fed.

Uses of Copper: Copper is used to treat cataracts (page 202) and osteoarthritis (page 176).

Dosage Information: The National Research Council recommends that adults consume from 1.5 to 3.0

milligrams of copper per day. Most people get sufficient copper in their diet and a multivitamin–mineral supplement. If you need additional copper because you are taking zinc supplements, take 1.5 to 3 milligrams daily with food. Typically, the ratio of zinc to copper is 10:1; in other words, if you're taking 30 milligrams of zinc per day, you would also take 3 milligrams of copper.

Possible Side Effects: When taken at recommended dosages, no side effects are expected. At high doses (10 milligrams or more), nausea, vomiting, muscle pain, and stomach pain may occur. Some experts believe excessive copper may be linked with autism and hyperactivity. Excessive copper may also cause damage to joint tissues.

Possible Interactions: People with Wilson's disease (a rare genetic disorder characterized by high copper levels) should not take copper supplements. If your drinking water travels through copper pipes, check the copper content of your water before taking supplements. Women who are pregnant or who are taking birth control pills should ask their physicians before taking copper supplements.

Folic Acid

Folic acid—also known as vitamin B9, folate, and folacin—plays many crucial roles in maintaining health. Folic acid works with vitamin B12 to create red blood cells. In fact, folic acid deficiency can lead to a particular form of anemia called megaloblastic anemia, after the technical name for improperly formed blood cells.

In addition, folic acid works with vitamin B12 to facilitate normal cell division and synthesize RNA and DNA, the genetic blueprints of every cell in the body. Women who are planning to become or who are pregnant need adequate levels of folic acid because it is instrumental in preventing most neural-tube birth defects as well as congenital abnormalities. Numerous studies have shown the benefit of folic acid supplementation throughout pregnancy to help prevent a birth defect known as spina bifida.

Folic acid also helps prevent heart disease by lowering levels of the amino acid homocysteine. It also has a key role in keeping the skin, nails, nerves, mucous membranes, hair, and blood healthy.

Good Food Sources: Avocados, bran, beets, celery, fortified cereal, legumes, lentils, okra, broccoli, citrus fruits, liver, salmon, green leafy vegetables, nuts, orange juice, seeds, and pecans. Folic acid is also added to enriched breads, flours, corn meals, pastas, rice, and other grain products. (Folic acid is also manufactured by our intestinal bacteria.)

Signs of Deficiency: Signs of folic acid deficiency include impaired cell division, anemia, headache, loss of appetite, diarrhea, fatigue, paleness, insomnia, and an inflamed, red tongue. Folic acid deficiency is most likely to occur among people who have gastrointestinal or malabsorption disorders, women taking oral contraceptives, pregnant women not taking vitamin supplements, alcoholics, and teenagers who have a poor diet.

Uses of Folic Acid: Folic acid is used to treat anemia (page 168), arteriosclerosis (page 183), cancer (page 193), constipation (page 211), depression (page 215), diarrhea (page 224), gingivitis (page

242), gout (page 247), and heart attack and cardio-vascular disease (page 255).

Dosage Information: The adult RDA is 400 micro-grams for adults; pregnant and breast-feeding women should get 800 micrograms daily. Look for multivitamin–mineral supplements with folic acid, preferably in the form of 5-methyl-tetra-hydrofolate, because this is the most bioactive form.

The body needs folic acid to properly use vitamin B12. If you are deficient in vitamin B12, intake of 1,000 micrograms folic acid may be needed to treat anemia caused by the B12 deficiency. Consult a health professional who is knowledgeable in vita-min B12 deficiency anemia.

Possible Side Effects: Folic acid is considered safe. High dosages of folic acid may hide the symptoms of vitamin B12 deficiency. If you have any reason to suspect a B12 deficiency, consult with a knowledge-able health professional before starting a folic acid supplement program.

Possible Interactions: Antacids can interfere with folic acid absorption. Drug interactions can occur be-tween folic acid and some antibiotics and medica-tions for malaria and seizures. Use of oral contraceptives may increase the need for folic acid.

Iodine

Iodine is a trace element necessary for the normal function of the thyroid gland. Specifically, the thyroid gland uses iodine in the production of the thyroid hor-mone. Iodine deficiency can result in goiter (a condi-tion in which the thyroid becomes enlarged in an effort to compensate for the body's lack of iodine), as well as impaired mental and physical development in children. Iodine deficiency during pregnancy can cause miscarriage and increased risk of infant mortality.

Good Food Sources: Iodized salt, seafood, seaweed, fish liver oil.

Signs of Deficiency: Signs of iodine deficiency include thyroid enlargement. Iodine deficiency is rare in industrialized countries because iodine is added to table salt.

Uses of Iodine: Iodine is used to treat iodine deficiency. It is also used topically and as an antiseptic for cuts.

Dosage Information: The RDA is 150 micrograms for adults, 200 micrograms for pregnant and breast-feeding women. The therapeutic dose is 50 to 300 micrograms. Supplemental iodine is not recommended unless a person shows signs of iodine deficiency.

Possible Side Effects: Taking too much iodine (more than 1,500 micrograms per day) may inhibit thyroid hormone secretion. Iodine supplementation may contribute to acne in some cases.

Possible Interactions: People with hypothyroid disorder should avoid high-iodine foods. When taken in large amounts, some raw foods (Brussels sprouts, turnips, beets, cassava, cabbage, kale, peaches, spinach) can block the uptake of iodine into the thyroid.

Iron

Iron is a trace mineral found in the hemoglobin molecule of red blood cells, the part of the blood that carries oxygen from the lungs to the rest of the body. It is also found in myoglobin, the form of hemoglobin found in muscle tissue. Iron works with several enzymes required for energy production and protein metabolism.

Iron deficiency, the most common nutrient deficiency in the United States, leaves the body's tissues lacking in sufficient oxygen, which can result in iron-deficient anemia and fatigue. Most doctors recommend that pregnant women take a supplement containing iron since the increased demand for iron can rarely be met through diet alone during pregnancy. Many women's vitamin formulas include iron, which can be useful for menstruating women. After menopause, most women do not need extra iron.

Good Food Sources: Dietary sources of iron come in two forms: heme iron, found in animal sources such as chicken, red meat, eggs, liver, and seafood; and nonheme iron, found in whole grains, nuts, dried fruit, dark green vegetables, lentils, legumes, brewer's yeast, tofu, and fortified cereals. The body absorbs heme iron somewhat more easily than it does nonheme iron; however, if you eat nonheme iron along with heme iron foods or foods containing vitamin C, iron absorption greatly improves.

Signs of Deficiency: Signs of iron deficiency include fatigue, weakness, headaches, anemia, and intolerance of cold.

Uses of Iron: Iron is used to treat anemia (page 168).

Dosage Information: Take iron supplements only under a doctor's care. Iron is best absorbed when taken 30 minutes before a meal. The RDA for iron is 10 milligrams for adult men, 15 milligrams for adult women, 30 milligrams for pregnant women, and 15 milligrams for lactating women.

Iron deficiency can have many causes. Menstruation is a common cause for adult women. Intake of certain foods and drugs can contribute to iron deficiency, including coffee, tea, soy-based products, tetracycline, and antacids, as well as high doses of calcium, zinc, and manganese supplements. Some people have a greater need for iron, including individuals who have hemorrhoids, bleeding stomach ulcers, Crohn's disease, or other conditions that cause poor absorption of iron or abnormal blood loss. People who take aspirin routinely, vegetarians, and long-distance runners also often need to supplement with iron. People who fall into any of the above-mentioned categories are potential candidates for iron supplementation.

Look for a product that contains ferrous fumarate, ferrous peptinate, or iron glycinate, in liquid or tablets. These forms cause constipation and indigestion less often than other forms.

Possible Side Effects: Excessive intake of iron—whether the result of megadosing or from taking iron when you do not have a deficiency—can inhibit function of the immune system, interfere with the absorption of phosphorus, cause headache, constipation, fatigue, dizziness, and vomiting, damage the intestinal tract, and increase the risk of cirrhosis, cancer, and heart attack. Taking too much iron can be a problem for the one out of every 250 Americans who has a genetic condition called hemochromatosis, which causes the body to absorb twice as much iron from food and supplements as other people do.

Possible Interactions: Iron absorption increases when it is taken with vitamin C or vitamin A and is decreased by intake of caffeine, calcium, zinc, and high-fiber foods.

Warning: Iron poisoning is one of the most common causes of childhood poisoning deaths, primarily in young children who take their mothers' prenatal vitamins. As few as six high-potency iron tablets taken at one sitting can kill a small toddler. Take care to store vitamins with iron safely out of the reach of children.

Magnesium

Magnesium is involved in a number of crucial bodily functions, from the creation of bone to the beating of the heart and the balance of sugar in the bloodstream. It plays a role in a number of cellular processes, including the formation of bone, proteins, cells, and fatty acids. In addition, magnesium stimulates activity of B vitamins, assists in clotting of blood, relaxes the muscles, aids in metabolism of carbohydrates and minerals, helps the body maintain a regular heart rhythm, and plays a central role in the formation of ATP (adenosine triphosphate), the fuel on which the body runs.

Good Food Sources: Nuts, whole grains, wheat bran, dark green vegetables, brown rice, garlic, apples, bananas, apricots, beans, dairy products, meat, fish, oysters, and scallops.

Signs of Deficiency: Signs of magnesium deficiency include fatigue, muscle weakness, twitching, nervousness, depression, abnormal heart rhythm, and loss of appetite. People who are most likely to be magnesium deficient are those who take laxatives or potassium-depleting drugs, as well as people with diabetes, heart failure, or an alcohol abuse problem.

Uses of Magnesium: Magnesium is used to treat angina (page 171), anxiety (page 173), chronic fatigue syndrome (page 204), constipation (page 211), diabetes (page 219), fibromyalgia (page 236), glaucoma (page 245), heart attack and cardiovascular disease (page 255), insomnia (page 279), migraine headache (page 250), Parkinson's disease (page 299), and premenstrual syndrome (page 302).

Dosage Information: The adult RDA is 325 milligrams for adults and 450 milligrams for pregnant and breast-feeding women. Because magnesium can compete with other minerals for absorption, it is best to get your magnesium in a multivitamin–mineral supplement and to take it with a calcium supplement. (A 2:1 ratio of calcium to magnesium is recommended by most physicians.) Magnesium sulfate, magnesium gluconate, and a magnesium-protein complex are available for oral use. Dolomite, a magnesium and calcium carbonate complex mined from the ground, is not recommended as a source of magnesium because of the possible risk of lead contamination.

Possible Side Effects: Excessive magnesium, which can mean as little as 350 to 500 milligrams for some people, can cause diarrhea. People who have kidney disease should avoid magnesium supplements.

Possible Interactions: Magnesium works closely with calcium and with vitamins B6 and should be taken with these nutrients to maintain appropriate nutrient balance. Magnesium can also interact with muscle relaxants, diuretics, ulcer medications, and anticoagulant drugs. Do not take magnesium supplements if you suffer from kidney disorders, unless recommended by a physician.

Manganese

Manganese is a mineral essential for healthy bone, skin, connective tissue, nerves, and cartilage, and for the activation of the important antioxidant enzyme superoxide dismutase (SOD). SOD helps prevent inflammation and free radical damage to the cells. Manganese is involved in many enzyme reactions, including those responsible for controlling blood sugar levels. It assists in blood clotting, the production of energy from food, and in the synthesis of protein. It is also essential for nervous system function and fat and vitamin metabolism.

Good Food Sources: Nuts, wheat bran, avocados, leafy green vegetables, pineapple, dried fruits, coffee, tea, and seeds.

Signs of Deficiency: Signs of manganese deficiency include cartilage problems, infertility, problems with fat and carbohydrate metabolism, and birth defects. Serious manganese deficiencies are rare. People with osteoporosis usually have low blood levels of manganese and can benefit from supplementation, as can people with diabetes.

Uses of Manganese: Manganese is used to treat diabetes (page 219) and ear infections (page 231).

Dosage Information: While the government has not established an RDA for manganese, the estimated minimum daily requirement for manganese is 2.5 to 5 milligrams. Most people do not consume enough to fall within that range. A multivitamin–mineral supplement that contains manganese is sufficient for most people. If you need a supplement, look for manganese citrate tablets or capsules.

Possible Side Effects: Manganese is very safe at the

levels found in supplements. People with cirrhosis of the liver should avoid manganese supplements because they may not be able to properly excrete this mineral.

Possible Interactions: Manganese works with copper and zinc to activate SOD. Both calcium and iron reduce the amount of manganese the body can absorb. Antacids and antiulcer drugs may interfere with the absorption of manganese.

Molybdenum

Molybdenum is a trace mineral essential for the formation of uric acid, a form of waste that is excreted as urine. If the body has too much molybdenum, it produces too much uric acid; if the body has too little molybdenum, it produces too little uric acid. When levels of uric acid exceed the amount the kidneys can process, it collects in the blood and settles in the joints, where it crystallizes and causes gout. In addition, molybdenum is essential for the utilization of iron, metabolism of carbohydrates, and the detoxification of sulfites.

Good Food Sources: Whole grains, meats, legumes. *Note*: Actual levels of molybdenum in these foods varies, depending on the mineral content of the soil in the region where the food is produced.

Signs of Deficiency: Molybdenum deficiency is exceedingly rare; possible symptoms include rapid heartbeat and night blindness.

Uses of Molybdenum: Molybdenum is used to treat gout (page 247).

Dosage Information: The government has not established an RDA for molybdenum. A safe and adequate range is considered 100 to 500 milligrams per day. Molybdenum is commercially available as sodium molybdate.

Possible Side Effects: Excessive levels of molybdenum (levels of 5,000 to 10,000 milligrams per day) have been known to cause gout.

Possible Interactions: None known.

Phosphorus

Phosphorus is the second most abundant mineral in the body (calcium is the first); phosphorus works with calcium to build and maintain bones and teeth. As the two most common minerals in the body, calcium and phosphorus depend on each other and must maintain a stable ratio (1:2 is ideal) to keep the body healthy. Phosphorus is found in all cells and is a key factor in the growth and maintenance of cells and tissues and in energy production. It also helps in the formation of cell membranes, DNA, and RNA.

Good Food Sources: Meat, poultry, fish, dairy products, nuts, legumes, beans, and grains.

Signs of Deficiency: Signs of phosphorus deficiency include fatigue, loss of appetite, weakness, bone pain, reduced bone mineralization, and muscle tremors. People most likely to be deficient in phosphorus are those who take large amounts of aluminum-containing antacids, and people who have kidney or liver disorders, any condition that hinders metabolism of vitamin D, or alcoholism.

Uses of Phosphorus: Phosphorus is used to treat osteoporosis (page 296) and phosphorus deficiency.

Dosage Information: The RDA for phosphorus is 800 milligrams for adults and 1,200 milligrams for pregnant or nursing women. Most people have no need to take phosphorus supplements, which are available in tablets and capsules. In fact, many people actually consume too much phosphorus, given that one serving of most soft drinks supplies up to 500 milligrams and that many convenience foods contain phosphoric acid as a preservative.

Possible Side Effects: Phosphorus supplements are not known to cause side effects at recommended levels. Excessive phosphorus, on the other hand, can cause calcium loss and osteoporosis.

Possible Interactions: None known.

Potassium

Potassium is an electrolyte necessary for the maintenance of regular heart rhythm, blood pressure, neuromuscular functioning, acid levels, and water balance. Levels of potassium tend to be low in people who are taking diuretics ("water pills") or laxatives, as well as those who have chronic diarrhea or kidney disorders.

Sodium and potassium surround cells as positively charged "ions." Together, sodium and potassium provide the electrical potential necessary for cell membranes to control the transfer of water into and out of the cell. For optimal health, the body needs to maintain a balance between sodium and potassium at a ratio of approximately 5:1, meaning a daily intake of five times as much potassium as sodium.

Good Food Sources: Bananas, apricots, figs, beans, garlic, brown rice, nuts, orange juice, potatoes, raisins, winter squash, peanut butter, and yams.

Signs of Deficiency: Signs of potassium deficiency include weakness, confusion, irritability, and problems with muscle contractions.

Uses of Potassium: Potassium is used to treat hypertension (page 266).

Dosage Information: The RDA for potassium is 900 milligrams for adults. Over-the-counter potassium is available in 99-milligram tablets, timed-release tablets, effervescent tablets, and capsules. (The federal government restricts supplements to 99 milligrams to reduce the risk of side effects.) Potassium supplements are sold as potassium salts (potassium chloride and potassium bicarbonate) or potassium bound to different mineral chelates (e.g., aspirate, citrate). Potassium supplements should be taken under supervision of a physician.

In general, dietary potassium is preferred over supplements to minimize the risk of unwanted side effects. Studies show that a diet low in potassium and high in sodium (salt) increases the risk of heart disease, stroke, and high blood pressure. This imbalance is particularly dangerous if you have kidney disease or high blood pressure or if you are taking certain medications, such as ACE inhibitors or potassium-sparing drugs.

Possible Side Effects: If you suffer from a kidney disorder, do not take potassium supplements without your doctor's recommendation. Potassium supplements can irritate the stomach, causing nausea, vomiting, and ulcers, if taken in amounts greater than 99 milligrams. The potassium available in fruit (a banana has 500 milligrams, for example) does not cause this side effect.

Possible Interactions: To minimize the risk of potassium toxicity, potassium supplements should not be taken along with ACE inhibitors, potassium-sparing diuretics, and many types of antibiotics.

Selenium

Selenium is a trace mineral that is believed to be a potent protector against cancer because selenium has been shown to activate the very powerful antioxidant enzyme glutathione peroxidase, an antioxidant found in every cell that works with vitamin E to prevent free-radical damage. Selenium also stimulates the thyroid hormones, prevents buildup of fats in the blood vessels, enhances immune-system functioning, and protects against heavy-metal poisoning.

It's been shown in more than twenty countries that the lower the intake of selenium, the higher the incidence of cancer of the colon, breast, pancreas, ovary, bladder, prostate, rectum, skin, and lungs. Similarly, low selenium intake has been associated with increased risk of cardiovascular disease, inflammatory diseases, cataracts, and premature aging.

Good Food Sources: Whole grains, asparagus, garlic, mushrooms, soybeans, tuna, seafood, pineapples, and brown rice. (The amount of selenium in the soil varies greatly by geographic region, and it affects the selenium level of food grown in the soil.)

Signs of Deficiency: Signs of selenium deficiency include dry scalp and skin problems.

Uses of Selenium: Selenium is used to treat allergies and asthma (page 158), anxiety (page 173), arteriosclerosis (page 183), cancer (page 193), cataracts (page 202), gingivitis (page 242), heart attack and cardiovascular disease (page 255), macular degeneration (page 286), and rheumatoid arthritis (page 180).

Dosage Information: The adult RDA is 50 micrograms; the typical therapeutic dose is 200 micrograms. Do not take higher doses unless you are under a doctor's care. Look for capsules and softgels, the latter of which are usually in combination with vitamin E. Some nutritionists say that a natural form of selenium, called L-selenomethionine or selenium-rich yeast, is superior to synthetic forms, but this has not been proven.

Possible Side Effects: Taking excessive amounts of selenium (1,000 micrograms or more) can cause rash, changes in the nervous system, and loss of fingernails.

Possible Interactions: Selenium supplements may increase the body's response to some chemotherapy drugs. High intake of vitamin C and zinc can interfere with selenium absorption.

Vitamin A (Beta-carotene)

Vitamin A and beta-carotene are antioxidants, which may help protect against cancer and improve resistance to certain diseases. They also help form and maintain healthy function of the eyes, hair, teeth, gums, and mucous membranes. In addition, vitamin A is involved in fat metabolism and the production of white blood cells.

Vitamin A is a fat-soluble vitamin that comes in two forms: retinol, found in animal tissues, and beta-carotene, found in plants. (Beta-carotene is sometimes called a provitamin because it must be broken down by the body into vitamin A before it acts as a vitamin.)

Good Food Sources: Whole milk, butter, organ meats, carrots, sweet potatoes, kale, butternut squash, spinach, arugula, red bell peppers, dark green vegetables, liver, cheese, fish liver oil, egg yolks, and apricots.

Signs of Deficiency: Signs of vitamin A deficiency include night blindness, slow or stunted growth in children, dry skin and eyes, increased susceptibility to infectious disease.

Uses of Vitamin A and Beta-Carotene: Vitamin A and beta-carotene are used to treat arteriosclerosis (page 183), bronchitis (page 188), cancer (page 193), cataracts (page 202), colds and flu (page 207), gingivitis (page 242), herpes (page 263), impotence (page 270), macular degeneration (page 286), osteoarthritis (page 296), and psoriasis (page 306).

Dosage Information: Vitamin A used to be labeled in IUs (International Units) but is now expressed in retinol equivalents (RE) to better distinguish between the two forms of vitamin A. Vitamin A: Women who are pregnant or could become pregnant should limit their intake of vitamin A to less than 10,000 IU (2,000 RE); men and postmenopausal women can take up to 25,000 IU (5,000 RE) daily. Beta-carotene: Adults can take up to 25,000 IU of beta-carotene (5,000 RE). Beta-carotene is preferred over vitamin A, and it is recommended that you buy a high-quality multivitamin–mineral supplement that contains beta-carotene. Some supplements state "Vitamin A (as beta-carotene)."

Vitamin A is available in capsules, tablets, and liquid; beta-carotene in capsules and tablets. The liquid form of vitamin A may be taken by drops directly into the mouth or mixed into juice or food.

Many cereals, juices, dairy products, and other processed foods are fortified with vitamin A.

Possible Side Effects: Intake of 25,000 IU or more per day of vitamin A can cause headache, hair loss, fatigue, bone problems, dry skin, and liver damage. Beta-carotene does not cause these problems, although taking more than 100,000 IU daily can cause the skin to have a yellow-orange hue. Use of vitamin A should be limited to less than 10,000 IU daily for women who are pregnant or who could become pregnant, because there is an increased risk of birth defects. In addition, vitamin A should not be taken in large amounts by pregnant women and people suffering from liver disease, diabetes, or hypothyroidism.

Possible Interactions: Taking vitamin A supplements while using prescription drugs derived from vitamin A (such as Accutane [isotretinoin] or Tegison [etretinate]) can result in toxic side effects.

Vitamin B Complex

Vitamin B-complex supplements contain the eight essential B vitamins in one tablet or capsule. Each of the B vitamins has its own chemical makeup, yet they perform similar functions and are found in many of the same foods. Some of the functions they share include maintaining healthy muscles and skin, enhancing the immune and nervous systems, promoting metabolism, and stimulating cell reproduction and growth.

Good Food Sources: Brewer's yeast, beans, peas, dark green leafy vegetables, whole-grain cereals, organ meats, and dairy products.

Signs of Deficiency: Deficiency of one of the B vitamins suggests a deficiency in the others as well. People who are susceptible to a vitamin B deficiency include alcoholics, people who eat a lot of sugar, the elderly, people with malabsorption conditions or who take antibiotics for a long time, pregnant women, and nursing mothers. Signs of deficiency include scaly, oily skin, stomach distress, headache, anxiety, moodiness, and heart arrhythmias.

Uses of B Complex Vitamins: B vitamin complex is used to treat canker sores (page 198), carpal tunnel syndrome (page 200), dandruff (page 214), diverticulitis (page 226), glaucoma (page 245), gout (page 247), heart attack and cardiovascular disease (page 255), irritable bowel syndrome (page 282), Parkinson's disease (page 299), and psoriasis (page 306). A deficiency of one B vitamin usually suggests low levels of others as well. For specific uses of B vitamins, see the individual listings below for each of the B vitamins.

Dosage Information: Tablets or capsules, often sold as B-50's and B-100's, which means the supplement supplies either 50 percent or 100 percent, respectively, of the DRI for each B vitamin. Take with food. If you are taking a multivitamin–mineral supplement, you probably do not need a B complex unless you are treating a specific condition.

Possible Side Effects: Side effects are rare and are usually seen only when the supplement is taken in extremely large amounts. Magnesium supplements can reduce absorption of B vitamins. You may need to increase your B vitamin intake when taking magnesium.

Possible Interactions: See the listings under the individual B vitamins below.

Vitamin B1 (Thiamin)

Thiamin, or vitamin B1, is a member of the B vitamin complex. Although the body needs only a miniscule amount of thiamin, it plays several major roles in health. Thiamin assists in carbohydrate metabolism and blood formation, stimulates blood circulation, and has a part in maintaining muscle tone of the stomach, intestines, and heart. Vitamin B1 is essential for healthy brain and nerve cell function, and it promotes appetite.

Good Food Sources: Dried beans, oatmeal, brown rice, peanuts, peas, soybeans, wheat germ, lean meats, fish, cereals, fortified breads, and whole grains.

Signs of Deficiency: Signs of thiamin deficiency include shortness of breath, low blood pressure, irregular heart rhythm, fatigue, nerve damage, anxiety, muscle cramps, and chest pain. Low thiamin levels can also cause beriberi, a nervous-system disorder in which people experience fatigue, weight loss, gastrointestinal disorders, weakness, and tender, atrophied muscles. A thiamin deficiency can be caused by alcohol abuse and lead to significant memory impairment, problems with motor and eye movements, and poor reality perception. Other people who may have an increased need for thiamin include pregnant women and people who exercise strenuously.

Uses of Thiamin: Thiamin is used to treat heart attack and cardiovascular disease (page 255).

Dosage Information: The adult RDA is 1.5 milligrams. The therapeutic dose is 1.5 milligrams daily. A high-quality multivitamin–mineral supplement should contain sufficient thiamin, usually as thiamin hydrochloride. It is also included in a vitamin B-complex supplement, as well as an individual nutrient in tablet and capsule form.

Possible Side Effects: Thiamin is safe when taken as directed. At levels of about 5 milligrams, thiamin can occasionally cause side effects, including itching, nervousness, flushing, and an abnormally rapid heartbeat (tachycardia) in sensitive individuals.

Possible Interactions: Certain medications, such as antibiotics, epileptic drugs, oral contraceptives, and sulfa drugs, can decrease thiamin levels in the body. When thiamin is taken as a supplement it is usually to prevent a deficiency or to treat impaired mental function in the elderly or people with Alzheimer's disease.

Vitamin B2 (Riboflavin)

Vitamin B2, or riboflavin, is part of the complex of water-soluble B vitamins. It plays a primary role in processing amino acids and fats, forming red blood cells, converting carbohydrates into energy, activating vitamin B6, and folic acid, and maintaining the mucous membranes in the digestive tract.

Good Food Sources: Legumes, brewer's yeast, soybean products, kidney, liver, milk, broccoli, Brussels sprouts, asparagus, eggs, legumes, spinach, yogurt, and meat.

Signs of Deficiency: Signs of riboflavin deficiency include depression; sores and cracks at the corners of the mouth; oily, dry, scaly skin; sensitivity to light; and swollen, red painful tongue.

Uses of Riboflavin: Riboflavin is used to treat cataracts (page 202).

Dosage Information: The adult RDA is 1.3 milligrams (females) and 1.8 milligrams (males). The therapeutic dose is 2 milligrams daily. The amount of riboflavin in good-quality multivitamin–mineral supplements or B-complex supplements is sufficient for most people. Look on the label for activated riboflavin (riboflav-5-phosphate) or simply riboflavin. It is also available as a sole nutrient in tablet form. Take with food to improve absorption. People who have an increased need for riboflavin include women who take oral contraceptives and anyone who participates in routine strenuous activity. Pregnant women need to be particularly careful to consume enough riboflavin, because a deficiency can cause damage to the fetus.

Possible Side Effects: Riboflavin may cause the urine to turn a dark yellow, but this is a completely harmless side effect.

Possible Interactions: Certain chemotherapy drugs can interfere with riboflavin metabolism.

Vitamin B3 (Niacin)

Niacin and niacinamide are the two main forms of vitamin B3. Both substances are key in releasing energy from carbohydrates, processing alcohol, forming fats, and producing sex hormones. A significant benefit of niacin is its ability to prevent recurrent heart attack. Niacin also helps regulate cholesterol levels. A third form of niacin, inositol hexaniacinate, is gaining acceptance as a substitute for niacin. Inositol hexaniacinate is composed of one molecule of inositol (an "unofficial" B vitamin) and six molecules of niacin.

Good Food Sources: Peanuts, brewer's yeast, fish, and whole grains.

Signs of Deficiency: Signs of niacin deficiency include fatigue, irritability, insomnia, blood sugar fluctuations, arthritis, diarrhea, loss of appetite, an inflamed tongue, and digestive problems. Niacin deficiency is rare in western cultures, largely because niacin is added to white flour, and Americans consume a great deal of these products.

Uses of Niacin: Niacin is used to treat depression (page 215).

Dosage Information: The adult RDA is 15 milligrams (for women) and 20 milligrams for men. The therapeutic dose is 18 milligrams. A multivitamin–mineral or B-complex supplement that contains niacin and/or niacinamide is sufficient for most people. Both forms of vitamin B3 are available individually as tablets and capsules. Niacinamide is preferred by many individuals because it does not cause the side effects associated with niacin. If you prefer to take niacin or your physician has approved it for you, do not buy sustained-release or slow-release niacin products, because they can be very harmful to the liver. Take niacin with food to reduce the chance of stomach upset.

Possible Side Effects: Flushing, a feeling of heat of the face and sometimes the entire body, nausea, and itching are associated with niacin. These side effects are temporary, lasting several minutes to about one hour. More serious effects include dark urine, yellow skin or eyes, and loss of appetite. If you have liver disease or low blood pressure, do not take niacin. Use of niacin or niacinamide may cause any of the following conditions to become worse: diabetes, glaucoma, gout, bleeding disorders, or stomach ulcer.

Possible Interactions: Niacin can decrease the effectiveness of insulin in diabetics and increase the effects of antihypertensive drugs in people with high blood pressure.

Vitamin B5 (Pantothenic acid)

Pantothenic acid, or vitamin B5, is part of the vitamin B complex. It has several critical roles in the body, including helping to convert proteins, carbohydrates, and fats into energy and aiding in the production of hormones and antibodies. It is referred to as an "anti-stress" vitamin by many experts because of its ability to relieve depression and anxiety. Pantothenic acid works synergistically with vitamin B1, B2, and B3 to make fuel for the body in the form of ATP.

Good Food Sources: Pantothenic acid can be found in almost all foods (pantothenic is derived from a Greek word that means "from everywhere"). The best sources include brewer's yeast, wheat germ, wheat bran, peanuts, peas, whole grains, broccoli, mushrooms, and sweet potatoes. Deficiency is rare, although people who abuse alcohol are likely to have low levels and are the best candidates for supplementation.

Signs of Deficiency: Signs of pantothenic acid deficiency include nausea, numbness of the extremities, muscle cramps, and stomach pain.

Uses of Pantothenic Acid: Pantothenic acid is used to

treat anemia (page 168), anxiety (page 173), and rheumatoid arthritis (page 180).

Dosage Information: If you take pantothenic acid as a separate supplement, take it along with a B complex supplement. A multivitamin–mineral supplement that contains either d-calcium pantotheate or pantothenic acid is adequate for most people. For additional amounts, pantothenic acid is available in capsules and tablets and in extended-release form.

Possibe Side Effects: Pantothenic acid is safe at suggested supplemental doses but may cause diarrhea if you take several grams per day.

Possible Interactions: None known.

Vitamin B6 (Pyridoxine, Pyridoxal, Pyridoxamine)

Vitamin B6 or pyridoxine is one of the water-soluble B-complex vitamins. Its main functions in the body are to help release energy from food (metabolism), aid in the proper functioning of more than sixty enzymes, promote a healthy immune system, help in cell multiplication, and assist in the manufacture of genetic material called nucleic acid. Large concentrations of pyridoxine are found in the brain, which has led to its use in treating depression. Vitamin B6 helps maintain brain function; aids in the formation of red blood cells; is essential to protein metabolism and absorption; and helps with the synthesis of antibodies in the immune system.

Good Food Sources: Meats, fish, nuts, legumes, bananas, brown rice, avocados, whole grains, lentils, corn, eggs, fortified cereals, spinach, potatoes, soybeans, liver, kidney, poultry, oatmeal, and prunes.

Signs of Pyridoxine Deficiency: Signs of pyridoxine deficiency usually occur with other deficiencies in the B complex; symptoms include weakness, inflamed tongue and mouth, sleeplessness, depression, confusion, inflammation of the mucous membranes in the mouth, and nerve problems in the feet and hands.

Uses of Pyridoxine: Pyridoxine is used to treat acne (page 158), allergies and asthma (page 160), anemia (page 168), carpal tunnel syndrome (page 200), diabetes (page 219), fibrocystic breast disease (page 234), insomnia (page 279), nausea (page 291), Parkinson's disease (page 299), premenstrual syndrome (page 302), and prostate enlargement (page 305).

Dosage Information: The adult RDA is 1.5 milligrams for women and 2 milligrams for men. The therapeutic dose is 2.5 milligrams. The most common supplement dosage is 10 to 25 milligrams, but a physician may recommend 200 milligrams or more depending on your needs. Do not crush or chew the capsules or tablets. Look for pyridoxine available as pyridoxal-5-phosphate, which is more bioavailable than the other form, pyridoxine hydrochloride. Both forms are sold as tablets and capsules and in extended-release formulas. The pyridoxine hydrochloride form is sufficient as long as you get enough riboflavin and magnesium in your diet or in supplements.

Possible Side Effects: If taken in excessive amounts (200 milligrams or more per day) for a long period of time, pyridoxine may cause loss of sensation in the hands and feet and difficulty walking.

Possible Interactions: Pyridoxine increases the bioavailability of magnesium, so it is suggested that you take these nutrients together. If you are taking the drug levodopa, consult your physician before taking pyridoxine.

Vitamin B12 (Cobalamin)

Vitamin B12, sometimes referred to as cyanocobalamin, is a key component in cell formation and longevity, proper digestion, protein synthesis, absorption of food, and metabolism of fats and carbohydrates. It also helps maintain fertility, and, along with the other B vitamins, helps produce neurotransmitters, chemicals that facilitate communication between nerves. This latter function makes B12 helpful in the prevention and treatment of depression and other mood disorders. Vitamin B12 aids in the formation of red blood cells; helps maintain the central nervous system; and helps the body use folic acid.

Good Food Sources: Milk and milk products, eggs, meat, poultry, liver, oysters, shellfish, and other animal products.

Signs of Deficiency: Vitamin B12 deficiency causes anemia, which may be caused by inadequate consumption of B12 or an inability to absorb it properly. Malabsorption of vitamin B12 is common and may be caused by certain diseases, such as colitis or celiac disease, by an insufficient amount of stomach acid, abnormal bacterial growth in the intestines, or previous stomach or intestinal surgery. A deficiency of B12 can take many years to become apparent, because the body stores this vitamin—up to 10 milligrams at a time—and very little is excreted. Signs of deficiency, in addition to anemia, include memory loss, abnormal gait, nerve damage, decreased reflexes, hallucination, eye problems, and digestive disorders.

Uses of Vitamin B12: Vitamin B12 is used to treat Alzheimer's disease (page 164), anemia (page 168), and depression (page 215).

Dosage Information: The adult RDA is 3 micrograms. The therapeutic dose is 5 to 250 micrograms. Supplements are often recommended for the elderly, people with digestive disorders, and strict vegetarians. Sublingual tablets are preferred, because the nutrient is readily absorbed through mucous membranes in the mouth. Vitamin B12 is also available in regular and extended-release tablets. Injections can be obtained from your physicians. Treatment of specific ailments and severe deficiency often requires high dosages, which may be best treated by injections from a physician. For health maintenance most high-quality multivitamin–mineral and B-complex supplements contain sufficient amounts of Vitamin B12.

Possible Side Effects: Vitamin B12 can be taken at ten thousand times the DRI (2.0 to 2.6 micrograms) and not cause side effects.

Possible Interactions: People who take antigout or anticoagulant medication or potassium supplements may have a problem absorbing dietary B12 and need additional supplementation.

Vitamin C (Ascorbic acid)

Vitamin C, or ascorbic acid, is a water-soluble vitamin that is perhaps best known for its ability to help fight colds and flu. That's because vitamin C is a powerful antioxidant that neutralizes potentially harmful organisms and enhances the immune system. Vitamin C helps promote healthy teeth and gums, aids in the absorption of iron, helps wound healing, and strengthens blood vessel walls.

Good Food Sources: Citrus fruits, red bell peppers, kale, kiwi fruit, broccoli, Brussels sprouts, cauliflower, strawberries, red cabbage, cantaloupe, rose hips, spinach, tomatoes, green peppers, parsley, dark green leafy vegetables, and potatoes.

Signs of Vitamin C Deficiency: A deficiency of vitamin C can cause scurvy, a disease characterized by bleeding gums, loose teeth, anemia, joint tenderness and swelling, poor wound healing, dry skin, loss of appetite, frequent bruising, and weakness. Scurvy is very rare in the United States; marginally deficient levels of vitamin C, however, are sometimes seen among the elderly, hospitalized patients, and people on very restrictive diets. These individuals often are susceptible to infection and also have slow wound healing.

Uses of Vitamin C: Vitamin C is used to treat allergies and asthma (page 160), arteriosclerosis (page 183), bronchitis (page 188), bursitis (page 191), cancer (page 193), canker sores (page 198), cataracts (page 202), colds and flu (page 207), depression (page 215), diabetes (page 219), ear infections (page 228), gallstones (page 239), gingivitis (page 242), glaucoma (page 245), hemorrhoids (page 262), hypertension (page 266), infertility (page 277), irritable bowel syndrome (page 282), macular degeneration (page 286), Parkinson's disease (page 299), psoriasis (page 306), ulcers (page 309), urinary tract infections (page 312), and varicose veins (page 314).

Dosage Information: The adult RDA is 60 milligrams. The therapeutic dose is 300 to 3,000 milligrams. Recommended forms are tablets or capsules in 500- or 1000-milligram doses for ease in dosing. Most people who take vitamin C take between 500 and 4,000 milligrams daily, in divided doses. Because vitamin C is eliminated from the body two to three hours after taking it, time-release formulas are preferred. Avoid the chewable tablets, because they can erode the enamel on your teeth. It is also available in powder and syrup.

Factors that increase people's requirement for vitamin C include smoking, exposure to smoke or other toxic fumes, and the following conditions: burns, congestive heart disease, diarrhea, rheumatic fever, rheumatoid arthritis, trauma, surgery, and infection.

Possible Side Effects: At high doses (3,000 milligrams or more) some people experience diarrhea. If you or your health-care practitioner has decided you need a high dose of vitamin C, start at a low dose and increase gradually (increase 500 to 1000 milligrams every two days). If you experience diarrhea before you reach your target dosage, reduce your dosage the next day to the previous day's level and maintain that dosage. Vitamin C is essentially nontoxic: whatever the body can't use is excreted in the urine.

Possible Interactions: Vitamin C increases the absorption of iron and copper. It can also interfere with blood tests for vitamin B12, so notify your physician if you are taking supplemental vitamin C.

Vitamin D (Calciferol)

Vitamin D is a unique substance in that the body produces it when sunlight hits the skin. This vitamin stimulates the absorption of calcium and helps fight breast and colon cancer. It also helps the body maintain proper blood levels of calcium and phosphorus.

Good Food Sources: Cold-water fish, egg yolks, butter, and dark green leafy vegetables. In addition, vitamin D, in the form of vitamin D2 (or ergocalciferol) is often added to milk and other foods.

Signs of Deficiency: Vitamin D deficiency results in diseases that are characterized by soft, poorly formed bones—rickets in children and osteomalacia in adults. Both conditions are rare in the United States except among the elderly, who are more likely to get little or no exposure to sunlight because of ill health.

Uses of Vitamin D: Vitamin D is used to treat osteoporosis (page 296).

Dosage Information: The adult RDA and the therapeutic dose is 400 IU. Elderly people who do not get enough sunlight should take 400 IU of vitamin D daily. A good-quality multivitamin–mineral supplement usually contains an appropriate amount of vitamin D. As a single nutrient it is available in tablet and capsule. The form most often used in supplements is vitamin D2. The prescription form of vitamin D is called calcitriol and is about ten times more potent than vitamin D2.

Possible Side Effects: Of all the vitamins, vitamin D has the most potential to be toxic. High intake of vitamin D (more than 1,000 IU daily) can result in kidney stones and calcium deposits in the internal organs.

Possible Interactions: Corticosteroids can increase the body's demand for vitamin D.

Vitamin E (Tocopherol, Tocotrienol)

Vitamin E is a fat-soluble vitamin that has strong antioxidant properties. One of its primary tasks is to prevent oxidation, a chemical reaction that can cause illness, disease, and other harmful effects. Vitamin E also plays a major role in maintaining proper functioning of the muscles and nerves, it helps in the formation of red blood cells, and it assists in the utilization of vitamin K.

Recent studies show that vitamin E is a major factor in preventing heart problems by helping stop oxidation of cholesterol in the arteries. It appears to protect against certain cancers, provide relief of fibrocystic breast disease and PMS, and help maintain metabolic control in diabetes.

Good Food Sources: Avocados, whole-grain cereals, dark green leafy vegetables, poultry, eggs, seafood, seeds, nuts, wheat germ, asparagus, and various oils (sunflower, almond, wheat germ, and hazelnut). Most people do not get a sufficient amount of vitamin E from their diets.

Signs of Deficiency: Signs of vitamin E deficiency include dry skin, lethargy, inability to concentrate, staggering gait, loss of balance, and anemia. People most likely to experience symptoms of vitamin E deficiency are the elderly, people with chronic liver disease, and those on very low-fat diets.

Uses of Vitamin E: Vitamin E is used to treat acne (page 158), Alzheimer's disease (page 164), arteriosclerosis (page 183), bronchitis (page 188), cancer (page 193), cataracts (page 202), constipation (page 211), diabetes (page 219), fibrocystic breast disease (page 234), gallstones (page 239), gingivitis (page 242), heart attack and cardiovascular disease (page 255), hemorrhoids (page 262), infertility (page 277), irritable bowel syndrome (page 282), macular degeneration (page 286), menopausal complaints (page 287), Parkinson's disease (page 299), premenstrual syndrome (page 302), rheumatoid arthritis (page 180), and ulcers (page 309).

Dosage Information: The adult RDA is 30 IU daily. The therapeutic dose is 100 to 400 IU daily. Studies of vitamin E show that a level of at least 100 to 400 IU is recommended for health and prevention of disease. The preferred form is natural vitamin E, which is derived from soybean or wheat-germ oil. Natural vitamin E is better absorbed than the synthetic forms, which are made from purified petroleum oil. Look for d-alpha-tocopherol on the package. Natural vitamin E comes in oil-filled capsules and dry tablets in 200- or 400-IU doses.

Possible Side Effects: Vitamin E does not cause any known side effects except in extremely high doses. People with rheumatic heart disease, an overactive thyroid, diabetes, or high blood pressure should consult with their physician before taking vitamin E.

Possible Interactions: Vitamin E can enhance the effects of blood-thinning medications, such as warfarin (Coumadin).

Vitamin K (Phylloquinone)

Vitamin K is necessary for blood clotting and bone metabolism.

Good Food Sources: Spinach, leafy greens, oats, wheat bran, potatoes, cabbage, cauliflower, corn, and soybeans.

Signs of Deficiency: Signs of vitamin K deficiency include excessive bleeding, liver damage. (Deficiency is usually caused by an inability to absorb the vitamin, rather than an inadequate intake.)

Uses of Vitamin K: Vitamin K is used to treat osteoporosis (page 296).

Dosage Information: The general dose of vitamin K is 150 micrograms daily; the amount is typically found in multivitamin–mineral supplements.

Possible Side Effects: None known.

Possible Interactions: Vitamin K can reduce the effectiveness of anticlotting drugs, such as warfarin (Coumadin).

Zinc

Zinc is a mineral that is critical to the work of more than three hundred enzymes in the body. These enzymes assist in cell reproduction, maintain vision, enhance the immune system, maintain fertility, repair wounds, synthesize protein, and perform many other functions.

Good Food Sources: Oysters, pumpkin seeds, spinach,

beef, pecans, cashews, lamb, almonds, turkey, sunflower seeds, and wheat germ.

Signs of Deficiency: Signs of deficiency include slow wound healing, loss of appetite, white spots on fingernails, joint pain, recurrent infection, and acne. Signs of overdose include nausea, vomiting, impaired immunity, high cholesterol, and abdominal pain. Most Americans consume less than the recommended amount of zinc, but those who are most likely to be significantly deficient are alcoholics and people with chronic kidney disease, malabsorption conditions, or sickle cell anemia.

Uses of Zinc: Zinc is used to treat acne (page 158), Alzheimer's disease (page 164), athlete's foot (page 186), bronchitis (page 188), bursitis (page 191), cataracts (page 202), colds and flu (page 207), dandruff (page 214), ear infection (page 228), eczema (page 231), gout (page 247), herpes (page 263), impotence (page 270), infertility (page 277), irritable bowel syndrome (page 282), macular degeneration (page 286), osteoarthritis (page 176), osteoporosis (page 296), prostate enlargement (page 305), rheumatoid arthritis (page 180), ulcers (page 309), urinary tract infections (page 312), and varicose veins (page 314).

Dosage Information: The adult RDA is 15 milligrams and 20 milligrams for pregnant and breast-feeding women. The therapeutic dose is 30 to 45 milligrams. Zinc is available in tablets, lozenges, or capsules. Look for zinc in the form of zinc picolinate, zinc aspirate, or zinc chelate, all of which are the most easily absorbed.

Avoid eating the following foods within two hours of taking a zinc supplement, because the body will not absorb the nutrient: bran, high-fiber foods, foods high in phosphorus such as milk and poultry, whole-grain breads and cereals. If you take copper, iron, or phosphorus supplements, take them at least two hours before or after taking zinc.

Possible Side Effects: Zinc lozenges may cause mouth irritation, nausea, stomachache, and a bad taste in some individuals. Supplementing with more than 300 milligrams daily may impair the immune system.

Possible Interactions: Zinc may decrease the absorption of many antibiotics. The use of oral contraceptives and many diuretics can interfere with zinc absorption.

Mother Nature is a pharmacist, and the leaves, bark, berries, flowers, seeds, and roots of plants are her drugs. Herbal medicines, or botanicals, can be used in the treatment of almost every ailment, and they're often cheaper, safer, and more effective than synthetic drugs. Of course, herbal medicine cannot replace prescription and over-the-counter drugs, but they can complement conventional drug treatment in many cases.

Herbal medicines have a proven track record. Around the world, four out of five people use herbs as the basis of their medical care. European doctors often rely on herbal remedies, but herbs are not as widely used in the United States, in part because drug companies prefer to create synthetic medicines that can be patented and sold for a profit. The gifts of Mother Nature cannot be patented, so pharmaceutical companies prefer to produce and package synthetic drugs.

Despite the American preference for man-made medicines, about 25 percent of all prescription drugs in the U.S contain active ingredients that come from plants. In a number of cases, drugs contain active ingredients that have been synthesized from chemicals similar to those that occur naturally in plants. For example, the herb ephedra contains ephedrine, a decongestant, and many allergy and cold-relief medications contain pseudoephedrine, a synthetic version of ephedrine.

Both pharmaceutical companies and modern herbalists owe a debt of gratitude to ancient herbalists who experimented with different plants and carefully monitored usage outcomes. Based on experience, herbalists learned that some plants healed and others harmed. Today we understand the basics of biochemistry and why many herbal treatments work, but traditional herbal medicine is based on tens of thousands of years of trial and error. The tradition extends back before recorded history: A grave site of a Neanderthal man shows the use of yarrow, marsh mallow, and other healing herbs from 60,000 years ago.

ARE HERBS SAFE AND EFFECTIVE?

An alarming number of people hesitate before swallowing a couple of Tylenol but think nothing of taking a couple of capsules of an herbal remedy because they believe it is "natural" and therefore cannot harm them. While it's true that most herbs are safer than synthetic drugs and have fewer serious side effects, if used incorrectly, herbs can be as potent and as dangerous as prescription drugs. Some of the most toxic substances known come from plants. For example, the powerful rodent poison and central nervous system stimulant strychnine is derived from the plant *nux vomica*.

Unfortunately, herbal medicines can be misused easily. Unlike drugs, botanicals don't have to undergo rigorous testing or approval from the U.S. Food and Drug Administration (FDA) before they wind up in health food stores. (The FDA classifies most herbal remedies as foods or food additives, rather than drugs.) That means it's up to the individual to know about the safety and efficacy of herbal treatments before using them. It also means individuals should follow the package directions carefully when determining how much of an herbal treatment to take and whether it is safe for internal or external use. The absence of labeled warnings doesn't necessary mean an herbal product is safe under all circumstances, so you should proceed with caution.

PRACTICING HERBAL MEDICINE

Herbal medicines come in a number of forms.

- *Tinctures* are made by soaking an herb in a grain alcohol solution for a specified amount of time, usually several hours to several days, depending on the herb. The solution is then strained, yielding the tincture. (Tinctures also can be made using wine or apple cider vinegar instead of grain alcohol.)

- *Extracts* are made by distilling off some of the alcohol, leaving behind a fluid with a higher concentration of the active ingredient. Most extracts are formed using vacuum distillation or filtration techniques, which do not require the use of high temperatures.

- *Capsules* and *tablets* take the process one step further. They are solid extracts in which all the fluid is removed and the remaining concentrated solid is ground into granules or powders and shaped as capsules or tablets. If desired, the capsules can be opened and used to make infusions or compresses.

• *Teas* are made by steeping or soaking herbs in boiling water for five minutes or so, then straining off the loose herbs. Teas should be made with one teaspoon of dried herb or three teaspoons of fresh herbs per cup of water.

• *Infusions* are prepared like teas, but the herbs steep for 10 to 20 minutes, so the solution is darker, stronger, and usually more bitter.

• *Decoctions* are basically infusions that are boiled instead of steeped; they are usually used for hard or woody herbs that include bark or roots.

• *Pellets* are sugar-based granules that are usually taken sublingually.

UNDERSTANDING CONCENTRATIONS AND POTENCY

The strength or potency of an herbal extract is expressed as a concentration in many cases. For example, an extract with a 4:1 concentration has 1 unit of extract derived from 4 units of herb. A tincture is usually a 1:10 concentration (10 units of tincture came from 1 unit of herbs.) In general, a solid extract is at least four times as potent as an equal amount of fluid extract, and forty times as potent as a tincture, if they are produced from the same quality of herb.

MEDICINAL HERBS

Many herbs have healing properties that can be used to treat a variety of medical conditions. The listings in this chapter include common herbs used to treat the medical conditions described in this book. It includes their beneficial effects on the body, as well as any negative side effects. The listing cannot identify every reported side effect, but an effort has been made to highlight the most important. If you suspect you are experiencing a side effect from an herb, contact a doctor and stop using it. As noted, some herbs should be taken internally, and others should be used externally (topically) to treat various conditions. Always follow package instructions carefully.

TREATMENT TIPS

- If the herb has a strong and unpleasant taste, dilute it in water or apple juice.

- Always start with the lowest possible dose or frequency and give more if necessary.

- Look for side effects: Keep an eye out for nausea, diarrhea or headache within an hour or two of taking an herb.

- Don't use herbal ointments or topical treatments on broken skin, unless directed.

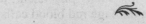

COMMON HERBS A TO Z

Alfalfa

The dried leaves of alfalfa (*Medicago sativa*) have been used for centuries to treat various disorders of the gastrointestinal tract, including loss of appetite and indigestion, among other conditions.

Alfalfa leaves contain saponins, which appear to prevent absorption of cholesterol. Other substances in the leaves include chlorophyll, flavones, isoflavones, sterols, protein, calcium, magnesium, and potassium. Overall, alfalfa leaves are thought to cleanse the body by acting as a laxative and diuretic and by stimulating the appetite.

Uses of Alfalfa: Alfalfa is used to treat flatulence (page 237), irritable bowel syndrome (page 282), and osteoarthritis (page 176).

Dosage Information: No therapeutic dose of alfalfa has been identified, but some experts recommend taking 1 to 2 milliliters of tincture daily. Powdered extract and capsules should be taken according to package directions. Tinctures, powdered extracts, and capsules are recommended.

Possible Side Effects: Diarrhea and stomach upset occur occasionally. If you develop these symptoms, stop taking the supplement immediately and consult your health-care provider. People with lupus or a history of the disease should avoid all alfalfa products. People with anemia should take alfalfa only with the permission of their physician, as the saponins in alfalfa can damage red blood cells.

Possible Interactions: If combining alfalfa with other herbs, you may need to reduce the dosage of alfalfa.

Aloe vera

Of the more than 120 types of aloe, the one most commonly grown and used is the aloe vera. The healing powers of aloe vera lie in its fleshy leaves, which contain a sticky gel and latex. The gel contains polysaccharides, which make it an effective external treatment for burns, cuts, and other skin irritations. When taken internally, the latex is effective against digestive problems.

Uses of Aloe Vera: Aloe vera is used to treat constipation (page 211), diverticulitis (page 226), gingivitis (page 242), and ulcers (page 309).

Dosage Information: Recommended forms are the bottled gel (juice), latex tablets, fluid extract, and lotion (for external use). It is also available as powder, powdered capsules, and softgels. Internal use of aloe should be done under the guidance of a physician. Topical use of the gel is effective for skin problems.

Possible Side Effects: Rash, diarrhea, or intestinal cramps are possible. Reduce the dose or stop taking the supplement. If you have a gastrointestinal condition, do not take aloe without first consulting your physician. Pregnant or lactating women can use aloe externally but should avoid using it internally. Do not exceed the recommended dose of aloe latex.

Possible Interactions: If you combine aloe with other herbs, you may need to reduce the dose.

Astragalus

More than two thousand types of astragalus are found around the world, but the most common form is Astragalus (*Astragalus membranaceus*), also known as milk vetch. This form contains polysaccharides, which experts believe are responsible for its healing abilities. Its effectiveness against immune-system conditions may be at least partially attributed to the high level of the antioxidant selenium in astragalus.

Uses of Astragalus: Astragalus is used to treat chronic fatigue syndrome (page 204) and colds and flu (page 207).

Dosage Information: This herb is available as a tincture, extract, capsules, and prepared tea bags. For dosage information, follow package directions.

Possible Side Effects: No side effects are noted. If you are pregnant, check with your doctor before taking astragalus.

Possible Interactions: Astragalus can clash with drugs designed to suppress the immune system (such as drugs used after organ transplant).

Bilberry

The bilberry (*Vaccinium myrtillus*), a relative of the blueberry, grows in the United States, Europe, and Canada. Bilberry helps prevent or heal fragile capillaries and other small blood vessels, which improves blood flow. These benefits appear to come from chemicals in the berries called anthocyanosides, which reportedly are up to fifty times more potent than vitamin E in antioxidant power.

Uses of Bilberry: Bilberry is used to treat cataracts

(page 202), glaucoma (page 245), hemorrhoids (page 262), macular degeneration (page 286), and varicose veins (page 314).

Dosage Information: Standardized forms are preferred; look for capsules, tinctures, or fluid extracts standardized to provide 25 percent anthocyanosides. Follow package directions for dosage information. You can also prepare an infusion by boiling 2 to 3 teaspoons of dried leaves in 1 cup of water. Take 1 cup per day. For dried berries, simmer 1 cup of water and 1 teaspoon dried berries for fifteen minutes. Drink 1 to 2 cups per day, cold.

Possible Side Effects: No side effects noted when taken as directed. Do not eat fresh bilberries, as they can cause diarrhea.

Possible Interactions: Bilberry does not interact with prescription drugs.

Black cohosh

Black cohosh (*Cimicifuga racemosa*) is a leafy, shrublike perennial that has been used for centuries to relieve menopause- and PMS-related pain and discomfort, especially hot flashes. Its healing power lies in its roots and rhizome, which contain an estrogenlike substance called formononetin. Formononetin helps reduce the secretion of luteinizing hormone, which is responsible for hot flashes and other menopausal symptoms. Black cohosh also has sedative qualities.

Uses of Black Cohosh: Black cohosh is used to treat incontinence (page 273), menopausal symptoms (page 287), and premenstrual syndrome (page 302).

Dosage Information: The standardized extract, powdered extract, and tincture are preferred. It is also

available as dried root or rhizome, capsules, syrup.
Follow package directions for dosage information.
To prepare a decoction from the dried root or rhizome, add ½ teaspoon to 1 cup of boiling water.
Cover and let steep and cool for thirty minutes. Take
2 tablespoons at a time as needed, up to 1 cup per
day, cold. Do not take black cohosh for longer than
six months.

Possible Side Effects: Contact your physician if you
experience any of these side effects: irritated uterus,
nausea, headache, drop in blood pressure or heart
rate, diarrhea, stomach pain, joint pain, breast tumors, or blood clots.

Warnings: Because serious side effects are possible,
this herb should be taken under medical supervision. Do not take black cohosh if you are pregnant,
have heart disease, are taking estrogen therapy, are
lactating, or if you have been advised not to take
oral contraceptives.

Possible Interactions: No interactions have been reported.

Boswellia

Also known as guggal, boswellia (*Boswellia serrata*)
is a tree that is grown in India. A gummy oleoresin,
also called guggal, is found in the tree trunk and consists of boswellic acids, which have antiinflammatory
properties. Boswellia has been a favorite antiarthritic
therapy among Ayurvedic practitioners for centuries.

Uses of Boswellia: Boswellia is used to treat bursitis
(page 191) and osteoarthritis (page 176).

Dosage Information: For topical use, buy the cream in
a base of *Boswellia serrata* and standardized for

boswellic acids. (Some creams also contain vitamin E.) Boswellia is also available as tablets and standardized extracts. Follow package directions for dosage information.

Possible Side Effects: Rare, but can include diarrhea, rash, and nausea. Avoid using if pregnant or breast-feeding.

Possible Interactions: None known.

Cayenne

Cayenne pepper (*Capsicum frutescens*) is the source of a resinous substance called capsaicin. When taken internally, capsaicin stimulates blood circulation, promotes sweating, and aids digestion. Topical applications are used to reduce inflammation and pain. Capsaicin provides temporary relief of pain by depleting small pain fibers of neurotransmitters called substance P. This substance is believed to be the main chemical mediator of pain messages. When the neurotransmitters are depleted, pain cannot be transmitted to the brain.

Uses of Cayenne: Cayenne is used to treat fibromyalgia (page 236).

Dosage Information: Look for ointment and lotion containing 0.025 to 0.075 percent capsaicin for external use; tincture for internal use. The ointment and lotion can be applied to unbroken skin. Gloves should be worn or the hands should be washed thoroughly after using capsaicin, because it can cause a burning sensation if it gets into the eyes, mouth, or nose. For dosage information for tincture, follow package information.

Possible Side Effects: Mild burning sensation may be felt at the application site or, in some cases, an allergic reaction may occur. Before the first treatment, apply a minute amount of capsaicin to a small area of skin to test for a possible negative reaction.

Possible Interactions: Cayenne may help block the ulcer-producing effects of aspirin and other nonsteroidal antiinflammatory agents.

Chamomile

Chamomile can refer to either the German or Hungarian form (*Matricaria chamomilla*) or to the English or Roman variety (*Anthemis nobilis*). Both species have similar medicinal properties, but the German chamomile is the most commonly used. The flowers, which contain bioflavonoids and volatile oils, provide the antiinflammatory and muscle-relaxing properties for which chamomile is known.

Uses of Chamomile: Chamomile is used to treat eczema (page 231), flatulence (page 237), gingivitis (page 242), and irritable bowel syndrome (page 282).

Dosage Information: The recommended and most common forms used are the dried flowers, from which an infusion can be made; prepared tea bags; and tincture. Chamomile is also available as tablets and capsules. To prepare an infusion, add 1 tablespoon of dried flowers (or 1 prepared tea bag) to 1 cup of boiling water. Allow to steep for ten to fifteen minutes. Drink 3 to 4 cups daily. For tincture and capsules, follow dosage information on the package.

Possible Side Effects: Side effects are rare. If you have allergies to ragweed or chrysanthemum, do not use chamomile. Chamomile is safe to use during pregnancy and lactation.

Possible Interactions: None known.

Chase tree berry

The chaste tree (*Vitex agnus-castus*), also known as vitex, is a small shrub that grows in Europe and in the southern United States. The berries of this shrub have the ability to restore the balance of estrogen and progesterone in women.

Uses of Chase Tree Berry: Chase tree berry is used to treat fibrocystic breast disease (page 234), infertility (page 277), menopausal symptoms (page 287), and premenstrual syndrome (page 302).

Dosage Information: The preferred forms are tincture, capsules (containing dried herb), and the berries. It is also available as a powder. To prepare an infusion using the berries, add 1 teaspoon ripe berries to 1 cup boiling water and let steep, covered, for ten to fifteen minutes. Drink 3 cups per day. For other forms, follow dosage information on the product package. Benefits are usually noticeable by day ten, but for best results, take chaste tree for six months or longer.

Possible Side Effects: Do not use this herb while using birth control pills. Stop taking chaste tree berry by the third month of pregnancy, as it stimulates breast milk production.

Possible Interactions: None known.

Dandelion

The dandelion (*Taraxacum officinale*), often regarded as a weed, has extremely high levels of vitamin A and good levels of vitamins C, D, and various B vitamins, plus iron, silicon, magnesium, manganese, and potassium. Women with PMS often take dandelion to help relieve water retention. The high level of potassium in the leaves makes dandelion the only potassium-sparing diuretic. (Diuretics typically deplete the body of potassium, which is the problem with most diuretic drugs.) The bitter components known as sesquiteropene lactones, found in both the roots and leaves, are responsible for dandelion's healing qualities.

Uses of Dandelion: Dandelion is used to treat canker sores (page 198), gallstones (page 239), menopausal symptoms (page 287), and premenstrual syndrome (page 302).

Dosage Information: Recommended forms include fluid extract, powdered solid extract, dried root, and dried leaves. It is also available as tablets, capsules, and tincture. The alcohol-based tincture should be avoided because an extremely high dosage is required to be effective. To prepare the dried root as a decoction, simmer 1 tablespoon in 8 ounces of boiling water for ten minutes; strain and drink up to 3 cups per day. For other forms, follow dosage information on the product package.

Possible Side Effects: When used to treat gallstones, gastritis, or stomach ulcer, dandelion should be used under supervision of a health-care professional. If you gather fresh dandelion leaves to dry for infusions, note that some people are allergic to the latex in the fresh leaves and experience a rash.

Possible Interactions: None known.

Dong quai

This Chinese herbal supplement, also known as angelic root (*Angelica sinensis*), has a long history of use for gynecological problems. Dong quai is believed to be effective because it balances women's hormone levels. It is also beneficial for people with high blood pressure and poor circulation. Dong quai is often mixed with other Chinese herbs to treat specific conditions.

Uses of Dong Quai: Dong quai is used to treat menopausal symptoms (page 287) and premenstrual syndrome (page 302).

Dosage Information: Fluid extract, tincture, and powdered root in capsules are the preferred forms; also available in tablets. The powdered root can be taken in capsules or used to make an infusion. Steep 1 to 2 grams in 8 ounces of boiling water for ten minutes. Drink 3 cups daily. For other forms, follow dosage information on the product package.

Possible Side Effects: Dong quai is safe when taken as directed. In large doses, however, it can cause diarrhea or abdominal bloating. Do not take during pregnancy.

Possible Interactions: None known.

Echinacea

Echinacea (*Echinacea purpurea*) is a wildflower also known as purple coneflower. It was widely used by Native Americans and was quickly adopted by the settlers for many uses, including common cold, syphilis, and snakebites. Echinacea contains several beneficial ingredients that are believed to have antibiotic and antiviral properties.

Uses of Echinacea: Echinacea is used to treat bronchitis (page 188), canker sores (page 198), colds and flu (page 207), ear infections (page 228), and herpes (page 263).

Dosage Information: Many experts believe the best

Echinacea product to buy is the fresh-pressed juice (not the same as a fluid extract) of *E. purpurea* standardized for a minimum of 2.4 percent beta 1,2-fructofuranosides. This form, however, is not always readily available. Other recommended forms include solid (dry powdered) extract standardized for 3.5 percent echinacoside, fluid extract, tincture, and freeze-dried plant. For dosage information, follow package directions.

Possible Side Effects: Generally, do not use echinacea for longer than two weeks. If you are treating a chronic immune-system condition, the usual recommendation is eight weeks of treatment followed by one week off. Consult with your physician before taking echinacea if you are pregnant or nursing. Combine echinacea with an equal amount of uva ursi to treat cystitis (urinary tract infections).

Possible Interactions: None known.

Evening primrose oil

Evening primrose oil is the best food source for an essential fatty acid called gammalinolenic acid (GLA). GLA converts to prostaglandin E1 (PGE1), a hormonelike substance that helps dilate blood vessels, thin the blood, and reduce inflammation. These properties help make evening primrose oil useful in the prevention of many hormone-related conditions.

Many people in the United States are deficient in GLA. Those most likely to have low levels of GLA include people with diabetes, eczema, and PMS, all of whom may have an inability to properly produce GLA.

Uses of Evening Primrose Oil: Evening primrose oil is used to treat acne (page 158), dandruff (page 214), eczema (page 231), fibrocystic breast disease (page 234), heart attack and cardiovascular disease (page 255), menopausal symptoms (page 237), premenstrual symptoms (page 302), psoriasis (page 306).

Dosage Information: Although no optimal intake level has been determined, experts generally recommend taking 3,000 to 6,000 milligrams of evening primrose oil daily, which supplies 270 to 360 milligrams of GLA. You need to take evening primrose oil for about thirty days before you will notice its benefits.

Possible Side Effects: Side effects are not common; headache, nausea, and rash are possible. These reactions usually disappear if you reduce the dosage. Evening primrose oil may increase mania in people with manic depression and exacerbate temporal lobe epilepsy. Women who have estrogen-related breast cancer should avoid this supplement because it promotes the production of estrogen.

Possible Interactions: To make prostaglandin E1, the body also requires magnesium, zinc, vitamin C, niacin, and vitamin B6. Some experts recommend taking these supplements along with evening primrose oil.

Fenugreek

Fenugreek (*Trigonella foenum-graecum*) is a member of the legume family and is one of the world's oldest medicinal herbs. In many parts of the world it is used as a food and a spice, as well as for medicinal purposes. The seeds of the fenugreek plant contain several substances that are responsible for its healing powers. One is mucilage, a gelatinous substance that is soothing to the digestive system and respiratory tract. Another is a group of alkaloids, which help in the treatment of diabetes by reducing the amount of sugar in the urine and improving glucose tolerance. Other components found in fenugreek include all of the B vitamins, vitamin A, and vitamin D.

Uses of Fenugreek: Fenugreek is used to treat diabetes (page 224) and menopausal symptoms (page 287).

Dosage Information: Fenugreek is available as powder, capsules, and tablets. Follow package directions for dosage information.

Possible Side Effects: Fenugreek is very safe when taken at recommended doses. If more than 100 grams per day is used, nausea and intestinal upset may occur. If uterine contractions occur, call your doctor immediately. If pregnant, do not take fenugreek because of possible uterine contractions.

Possible Interactions: If you combine fenugreek with other herbs, you may need to decrease the dosage.

Feverfew

Feverfew (*Tanacetum parthenium* or *Chrysanthemum parthenium*) is a flowering plant that has been used since ancient times to treat migraine, inflammation, and pain associated with menstruation. In the 1970s British researchers discovered that feverfew leaves are effective in the treatment of migraine. They attribute its pain-relieving ability to parthenolide. This chemical prevents substances that cause inflammation from entering the bloodstream and traveling to the blood vessels in the brain, where they appear to be at least partially responsible for migraine pain.

Uses of Feverfew: Feverfew is used to treat migraine (page 250).

Dosage Information: Feverfew is available as standardized extract, tincture, or capsule, as well as fresh leaves. Follow package information for dosage recommendations. Chewing fresh leaves (two to three medium-sized leaves per day) provides the best relief, but they are very bitter. The leaves can be put into a sandwich to make them more palatable. To make an infusion with the dried bulk herb, add 2 to 3 teaspoons dried herb to 8 ounces boiling water and allow to steep for five to ten minutes. Drink 2 to 3 cups daily.

Possible Side Effects: Chewing feverfew may cause sores in the mouth or stomach upset. Pregnant women should not take feverfew, because it may cause uterine contractions. Do not take feverfew if you have a blood-clotting disorder or are taking anticoagulant medication. Allow two to three months of taking feverfew daily before results become apparent.

Possible Interactions: None known.

Garlic

One of the oldest and most commonly used medicinal plants is garlic (*Allium sativum*). It contains sulfur compounds, including allicin, ajoene, allyl sulfides, and vinyldithins, which are responsible for garlic's antibiotic actions.

Uses of Garlic: Garlic is used to treat allergies and asthma (page 160), Alzheimer's disease (page 164), anxiety (page 173), arteriosclerosis (page 183), athlete's foot (page 186), bronchitis (page 188), cancer (page 193), colds and flu (page 207), diabetes (page 224), diarrhea (page 226), diverticulitis (page 228), ear infection (page 231), fibrocystic breast disease (page 234), gout (page 247), heart attack and cardiovascular disease (page 255), herpes (page 263), hypertension (page 266), and urinary tract infections (page 312).

Dosage Information: Look for products standardized to provide up to 5,000 micrograms allicin daily. To avoid smelling like garlic, odor-free, enteric-coated tablets and capsules are available; garlic oil and tinctures are also on the market. For dosage information, follow package recommendations.

Possible Side Effects: People who are sensitive to garlic may experience flatulence or heartburn. Garlic has anticlotting properties; therefore, if you are taking anticoagulant drugs, do not take garlic supplements without first consulting your physician.

Possible Interactions: Garlic may increase the potency of blood-thinning drugs, such as warfarin (Coumadin). Garlic may also increase the effectiveness of drugs used to lower blood sugar levels in the treatment of Type II diabetes.

Germanium

Germanium is a trace element (an essential substance found in minute levels in the body) found naturally in garlic, sea algae, aloe vera, shiitake mushrooms, Siberian ginseng, and barley. It is classified as an immunostimulant because it activates the immune system and promotes the production of gamma-interferon, an anticancer, antiviral substance. This naturally occurring element does not appear to have a significant impact on health, but synthetic organic compounds developed by Japanese researchers do. One of the most common of these compounds is called Ge-132, or organic germanium sesquioxide.

Organic germanium compounds fight viral infections, improve circulation, balance bodily functions, and boost the immune system. It also is a fast-acting painkiller and improves circulation to the brain. Germanium works by attaching itself to oxygen molecules, which deliver the mineral to the body's cells and enhance the oxygen supply. The increased oxygen allows the body to rid itself of poisons and toxins.

Uses of Germanium: Germanium is used to treat fibrocystic breast disease (page 234) and gout (page 247).

Dosage Information: The recommended form is colloidal liquid. It is also available in powder, granules, capsules, and tablets. Germanium is expensive and not as readily available as many other supplements. For dosage information, follow package directions.

Possible Side Effects: Minor skin eruptions, which disappear within a few days, occur in a small number of people. Loose stools may occur if high doses of germanium (more than 400 milligrams daily) are taken for more than thirty days.

Possible Interactions: None known.

Ginger

The dried rhizome of this perennial plant has been used for thousands of years for various gastrointestinal problems, such as bloating, loose stools, flatulence, nausea, and vomiting, and for treatment of inflammatory conditions, such as rheumatoid arthritis. Ginger (*Zingiber officinale*) is also effective in the treatment of menstrual cramps and migraine headache. The volatile oils in dried rhizome of ginger, including gingerols, shogaols, zingiberene, and bisablene, are responsible for its healing powers.

Uses of Ginger: Ginger is used to treat flatulence (page 237), heartburn (page 253), migraine headache (page 250), nausea and morning sickness (page 291), and osteoarthritis (page 176).

Dosage Information: Ginseng is available as dried root powder, fresh root, tablets, capsules, tincture, and prepared tea bags. The dried root powder can be added to liquids or food. To make a decoction, simmer 1 to 2 teaspoons dried root powder in 1 cup water for five to ten minutes; take as needed. To make a fresh root decoction, simmer 1 teaspoon grated fresh root in 1 cup water for fifteen minutes. Strain and take as needed. For other forms, follow package directions.

Possible Side Effects: Some people experience temporary heartburn. Long-term use of ginger during pregnancy is not recommended. If you have a history of gallstones, consult your physician before taking ginger.

Possible Interactions: If combining ginger with other herbs, you may need to reduce the amount of ginger.

Ginkgo

The fan-shaped leaves of the ginkgo tree (*Ginkgo biloba*) have been highly regarded for their medicinal value for thousands of years. Ginkgo is effective against vascular diseases and improves circulation, especially to the lower legs and feet and to the brain, where it may improve memory and concentration. Ginkgo contains various bioflavonoids, including quercetin and flavoglycosides, which appear to be the plant's healing compounds.

Uses of Ginkgo: Ginkgo is used to treat Alzheimer's disease (page 164), glaucoma (page 245), heart attack and cardiovascular disease (page 255), heartburn (page 253), impotence (page 270), and macular degeneration (page 286).

Dosage Information: Use a standardized extract of 24 percent flavoglycosides and 6 percent terpene lactones. It is also available in standardized capsules, tablets, and dry bulk. Do not buy nonstandardized products, as there is extreme variation in the content of its active ingredients. For dosage information, follow package recommendations.

Possible Side Effects: If you experience diarrhea, vomiting, nausea, irritability, or restlessness when taking ginkgo, consult your health-care practitioner to determine if you should reduce the dosage or stop taking the herb. Do not take ginkgo if you have a clotting disorder or if you are nursing or pregnant.

Possible Interactions: Ginkgo may increase the effect of blood-thinning drugs, such as warfarin (Coumadin).

Ginseng

Ginseng is an herb that is available in several species: Siberian ginseng (*Eleutherococcus senticosus*), American ginseng (*Panax quinquefolium*), Chinese or Korean ginseng (*Panax ginseng*), and Japanese ginseng (*Panax japonicum*). The most common and widely used is Chinese ginseng.

All forms of ginseng are often used to battle fatigue and increase energy levels; thus ginseng is usually regarded as a general tonic. Ginseng has the ability to regulate blood sugar fluctuations, and to stimulate circulation. When taken in lower doses it can increase blood pressure; at high doses it reduces it. The ability to adapt to the body's needs places ginseng in the category of "adaptogens," substances that regulate and normalize the body's systems. The active ingredients in American and Asian ginseng are known as ginsenosides, while those in Siberian ginseng are called eleutherosides.

Uses of Gingeng: Ginseng is used to treat chronic fatigue syndrome (page 204), diabetes (page 219), and impotence (page 270).

Dosage Information: Look for American and Asian ginseng products that are standardized for 5 to 9 percent ginsenosides, or Siberian ginseng standardized for more than 1 percent eleutherosides. Look for name brands with a reputation for quality. Suggested forms are any reputable standardized powdered or liquid extract, tincture, capsules, or fresh root. Also available are granules and tablets. Many manufacturers produce supplements that contain some ginseng as part of the formula; however, the levels are usually too low to have any therapeutic impact. Follow package recommendations for dosage information. If taking American or Asian ginseng for an extended time, follow each fifteen- to twenty-day course of treatment with a two-week notreatment period. For long-term treatment with Siberian ginseng, follow each sixty-day course with a two- to three-week no-treatment period.

Possible Side Effects: All forms of ginseng may cause the same types of side effects, including headache, insomnia, breast tenderness, anxiety, and rash. More serious reactions may include asthma attacks, increased blood pressure, and heart palpitations.

Possible Interactions: Ginseng may increase the effectiveness of insulin and drugs to lower blood-sugar levels.

Goldenseal

Goldenseal (*Hydrastis canadensis*) is an herb whose root and rhizome are valued for their antiinflammatory and antimicrobial properties. It also has the ability to aid digestion and dry up secretions.

The Native Americans used goldenseal to treat various types of inflammation and irritations of the urinary, respiratory, and digestive tracts. Today it is still used for similar conditions. Berberine, the main alkaloid found in goldenseal, is responsible for the plant's antimicrobial action, especially against *Escherichia coli*, *Salmonella typhi*, and the species *Chlamydia*. Other alkaloids, including hydrastine, berberastine, and canadine, also have some medicinal value.

Uses of Goldenseal: Goldenseal is used to treat athlete's foot (page 186), diarrhea (page 224), flatulence (page 237), gingivitis (page 242), herpes (page 263), and urinary-tract infections (page 312).

Dosage Information: Any forms standardized for 8 to 12 percent alkaloid content are recommended, including powdered dry extract, fluid extract, tincture, and capsules. You can also buy dried root to make decoctions. Follow package directions for dosage information. To make a decoction, pour 8 ounces of boiling water over 2 teaspoon goldenseal powder and steep covered for ten to fifteen minutes. Drink up to 3 cups daily. Do not use goldenseal for more than three weeks continuously. After three weeks, stop treatment for at least two weeks before starting again.

Possible Side Effects: In high doses, goldenseal may cause nausea, diarrhea, or irritation of the skin, throat, vagina, or mouth. Stop taking the herb if any of these reactions occurs. Do not take goldenseal if you are pregnant or nursing, or if you have diabetes, glaucoma, high blood pressure, heart disease, or if you have had a stroke.

Possible Interactions: None known.

Green tea

Green tea is derived from the same plant as are black and oolong teas, *Camellia sinensis*. The difference lies in how the tea leaves are prepared. Green tea maintains all of its active constituents because the leaves are not fermented, as are those of both black and oolong teas. These active components, called polyphenols, appear to be responsible for green tea's beneficial properties. These include the ability to protect against heart disease by lowering blood pressure, lowering total cholesterol levels, and reducing platelet aggregation (the clumping of blood cells that may lead to clotting). Green tea has also demonstrated the ability to reduce the risk of some cancers, especially esophageal cancer, and to fight bacteria, including the type that causes gingivitis and bad breath.

Uses of Green Tea: Green tea is used to treat cancer (page 193), eczema (page 231), gingivitis (page 242), and heart attack and cardiovascular disease (page 255).

Dosage Information: Look for capsules with an extract standardized for 80 percent total polyphenol and 55 percent epigallocatechin gallate (a type of polyphenol), or look for loose tea. (If you want to avoid the caffeine, purchase decaffeinated varieties.) To make tea, pour 8 ounces of boiling water over 1 teaspoon of tea leaves and steep for three minutes. Drink up to 5 cups daily, preferably with meals. For other forms, follow dosage recommendations on the package.

Possible Side Effects: Drinking large amounts of tea may cause insomnia, anxiety, or restlessness because of the caffeine content; green tea typically does not cause these symptoms. You may want to use a decaffeinated brand if you are sensitive to caffeine.

Possible Interactions: The effects of the caffeine may be exacerbated by birth control pills and various antibiotics.

Hawthorn

Hawthorn (*Crataegus laeviata,* also *Crataegus oxyacantha, Crataegus monogyna*) is a European shrub whose leaves, flower, and fruit are often used to treat heart problems, including angina, cardiac arrhythmias, heart palpitations, congestive heart failure, and atherosclerosis. The compounds present in hawthorn that give this herb its healing properties include quercetin, oligomeric procyanidine, and vitexin. These ingredients help treat high blood pressure, improve contractions of the heart, enhance coronary-artery blood flow, lower blood pressure, and reduce production of angiotensin II, a substance that constricts blood vessels.

Uses of Hawthorn: Hawthorn is used to treat angina (page 171), arteriosclerosis (page 183), heart attack and cardiovascular disease (page 255), and hypertension (page 266).

Dosage Information: The preferred form is the extract standardized for total bioflavonoid content of 2.2 percent or oligomeric procyanidins of 18.75 percent. Look for tincture, fluid extract, capsules, and dried leaves and berries. Follow dosage recommendations on product packages. To make an infusion of the dried leaves or berries, add 3 to 5 grams of herb to 8 ounces of boiling water and allow to steep ten minutes. Drink up to three cups daily. Always consult with your physician before starting hawthorn. Allow one to two months for the effects to be noticeable.

Possible Side Effects: Hawthorn is safe for long-term use and for use during pregnancy and nursing.

Possible Interactions: None known.

Licorice

Licorice (*Glycyrrhiza glabra*) is a sweet herb that is useful both for its medicinal qualities and as an ingredient in herbal formulas to mask any unpleasant tastes. It is a major herb in Chinese medicine, where it is used to treat conditions of the digestive and urinary tract. The active ingredients in licorice include glycyrrhizin and flavonoids. Glycyrrhizin is an antiinflammatory and antiviral substance, while the flavonoids are potent antioxidants, which help protect liver cells. Licorice acts as a mild laxative and also provides a protective coating to the stomach, which helps prevent stomach ulcers. When used for digestive disorders, a form of licorice called deglycyrrhizinated licorice, or DGL, is preferred, because the glycyrrhizin causes an increase in blood pressure in many people.

Uses of Licorice: Licorice is used to treat allergies and asthma (page 160), canker sores (page 198), chronic fatigue syndrome (page 204), colds and flu (page 207), constipation (page 211), eczema (page 231), heartburn (page 253), herpes (page 261), nausea (page 291), premenstrual syndrome (page 302), and ulcers (page 309).

Dosage Information: The recommended forms are dry powdered extract, fluid extract, and powdered root in capsules. Licorice is also available in chewable tablets, dried root, and lotion. Deglycyrrhizinated licorice is preferred for most cases, as glycyrrhizin causes side effects. Regular licorice, with glycyrrhizin, is indicated for respiratory infections and herpes. Follow dosage information on product labels.

Possible Side Effects: Mild cases of upset stomach, diarrhea, headache, edema, and grogginess may occur when taking either form of licorice. An increase in blood pressure and edema may occur in people who are susceptible to glycyrrhizin. This effect is usually seen in people who take more than 1 gram of glycyrrhizin daily for more than several weeks; therefore a DGL form is preferred. Avoid licorice if you have kidney failure, a history of high blood pressure, or if you are taking medication for heart problems. Consult your physician before starting licorice.

Possible Interactions: Licorice may enhance the effects of corticosteroid drugs.

Milk thistle

Milk thistle (*Silybum marianum*) has been used for more than two thousand years to treat liver problems and gallstones. The medicinal parts of the plant are the seeds and the dried flower, which contains silymarin. This bioflavonoid stimulates the liver to produce healthy liver cells and acts as an antioxidant to protect the liver from damage from free radicals.

Uses of Milk Thistle: Milk thistle is used to treat gallstones (page 239) and psoriasis (page 306).

Dosage Information: Look for the extract, standardized to 70 to 80 percent silymarin, found in capsules and as a tincture. Milk thistle is also available as dry bulk seeds. Follow dosage information on product labels. The dried, powdered seeds can be made into an infusion (2 to 3 teaspoons steeped in 8 ounces of hot water for ten to fifteen minutes) or added to food; 1 teaspoon three times daily.

Possible Side Effects: Milk thistle is considered very safe, even for women who are pregnant or lactating. It may cause loose stools in some people, but this effect usually disappears in two to three days.

Possible Interactions: None known.

Myrrh

Myrrh (*Commiphora molmol*) was a common remedy among ancient peoples for treatment of infections, dental problems, and bad breath. It is effective because it stimulates production of white blood cells and has strong antibacterial activity. Myrrh is an oil found in the bark of several different African and Arabian shrubs. The oil hardens into a resin, which is dried and powdered for use.

Uses of Myrrh: Myrrh is used to treat canker sores (page 198) and gingivitis (page 242).

Dosage Information: The preferred forms are tincture, powder, or capsule, depending on the use. Myrrh is also found in some tooth products. For dosage information, follow package directions. To make a tea, steep 1 to 2 teaspoons of powdered herb in 8 ounces of boiling water for ten to fifteen minutes. Drain, and drink three times daily. To make a mouthwash, steep 1 teaspoon powdered herb and 1 teaspoon boric acid in 16 ounces of boiling water. Let stand thirty minutes, strain, and use when cool.

Possible Side Effects: Mild stomach upset and diarrhea may occur. If myrrh is taken in large doses, it can cause vomiting, nausea, sweating, kidney problems, and heart palpitations. Do not take if you are pregnant, nursing, or have kidney disease.

Possible Interactions: None known.

Nettle

Nettle, or stinging nettle (*Urtica dioica*), gets its name from the fact that the tiny hairs on the leaves sting or burn when they touch the skin. When the plant is dried, however, it loses its stinging abilities. The medicinal parts of the plant include the root and leaves; the active substances in nettle are believed to be lectins (protein-sugar molecules) and polysaccharides (complex sugars). Nettle leaves prevent the body from producing prostaglandins, which cause inflammation, and the roots affect the sex hormones.

Uses of Nettle: Nettle is used to treat allergies and asthma (page 160), anemia (page 168), gout (page 247), prostate enlargement (page 305), and urinary-tract infections (page 312).

Dosage Information: The preferred forms are capsules, root extract, and dried root for decoctions. Also available are tablets, tincture, and dried whole herb. To prepare a decoction, steep 1 teaspoon dried root in 8 ounces of hot water and divide into two doses for the day.

Possible Side Effects: No side effects have been noted.

Possible Interactions: None known.

Oats

Oats (*Avena ativa*) are a widely used supplement for treatment of skin disorders such as eczema. The oats used in supplements are derived from wild species that are now cultivated worldwide, and they come in a variety of forms. Several parts of the plant are used medicinally. The seeds contain alkaloids, iron, manganese, and zinc; the straw has a high silica content. Alkaloids in the plant are believed to be responsible for the plant's ability to soothe and heal.

Uses of Oats: Oats are used to treat depression (page 215), eczema (page 231), and psoriasis (page 306).

Dosage Information: The recommended form for treatment of skin disorders is the dried herb, sometimes referred to as oat straw. Oats can be used externally in bathwater.

Possible Side Effects: No known side effects are noted when oats are used externally.

Possible Interactions: None known.

Peppermint

Peppermint (*Mentha piperita*) is actually a hybrid of two other mints—spearmint and water mint—and was first cultivated in 1750. Since it was first recognized for its healing powers, peppermint has been used to treat stomach and intestinal problems, including irritable bowel syndrome. The pure volatile oils, which contain menthol alone and menthol combined with other substances, appear to be the compounds that provide pain relief. They may also help tone the digestive system and promote the flow of bile from the gallbladder.

Uses of Peppermint: Peppermint is used to treat flatulence (page 237), gallstones (page 239), heartburn (page 253), irritable bowel syndrome (page 282), nausea and morning sickness (page 291).

Dosage Information: The preferred forms are dried leaves (for infusions), enteric-coated capsules (oil), and tincture. Peppermint is also available as tablets. To prepare an infusion, use 1 to 2 heaping teaspoons per 8 ounces of boiling water and steep for ten minutes. Drink up to 3 cups daily. For other forms, follow package directions.

Possible Side Effects: Do not ingest pure peppermint oil or pure menthol, as both are very toxic and, in the case of menthol, can be fatal. Those who should avoid using peppermint include women with a history of miscarriage and anyone with gallbladder or bile-duct inflammation, obstruction, or a related condition.

Possible Interactions: None known.

Psyllium

Psyllium is a flowering herb that produces small brown seed pods, which have been used for centuries to treat digestive problems. Psyllium seeds are a rich source of fiber, which is the ingredient that makes this herb so effective. It is also the reason many commercial laxatives contain psyllium as their main ingredient.

Uses of Psyllium: Psyllium is used to treat constipation (page 211), diverticulitis (page 226), gallstones (page 239), hemorrhoids (page 262), and ulcers (page 309).

Dosage Information: Psyllium is also found in many commercial bulk-forming laxative products. If you have been eating a low-fiber diet, begin gradually with ½ teaspoon of psyllium seeds or powder mixed in 8 ounces of cool liquid and drink 2 to 3 cups daily. Increase to 1 teaspoon per 8 ounces after one or two days. Stir the mixture vigorously and drink it quickly, followed by additional water. When taking psyllium it is important to drink at least eight to ten additional glasses of water daily to prevent blockage of the intestinal tract.

Possible Side Effects: People who are allergic to grasses or dust may have an allergic reaction to psyllium. Severe reactions are rare. Do not use psyllium if you are pregnant, because it can stimulate the lower pelvis.

Possible Interactions: None known.

Saw palmetto

Saw palmetto (*Serenoa repens*) is a shrub that develops berries that are used to treat a common problem among men, benign enlargement of the prostate gland. Saw palmetto is effective because it apparently blocks production of the chemical dihydrotestosterone, which is believed to cause this prostate problem. Urinary flow, which is often significantly reduced in men with prostate problems, can improve by 50 percent or more in men who take saw palmetto.

Uses of Saw Palmetto: Saw palmetto is used to treat prostate enlargement (page 305).

Dosage Information: The most effective form is the extract standardized to contain 85 to 95 percent fatty acids and sterols. Follow dosage information on product label.

Possible Side Effects: No side effects have been noted.

Possible Interactions: None known.

Skullcap

Skullcap (*Scutellaria baicalensis, S. lateriflora*) is a Chinese herb that has gained acceptability among Westerners, especially as a safe treatment for insomnia. Since it was first mentioned in Chinese writings back in A.D. 250–330, it has been valued as a diuretic, an antiinflammatory, and for its calming effects.

Uses of Skullcap: Skullcap is used to treat anxiety (page 173), headaches (page 250), and insomnia (page 279).

Dosage Information: The preferred forms are capsules, tincture, and dried leaves for infusions. Also available are tablets and the prepared tea. To prepare an infusion, pour 8 ounces boiling water over 2 teaspoons dried herb and steep covered for ten to fifteen minutes. Drink up to 3 cups daily. Skullcap is best taken after meals. For other forms, follow package information for dosage recommendations.

Possible Side Effects: Skullcap may cause drowsiness, so do not drive a car or operate heavy machinery after taking it. Other temporary side effects include stomach upset and diarrhea.

Possible Interactions: None known.

St. John's wort

St. John's wort (*Hypericum perforatum*) is a flowering herb that is taken internally to treat mild to moderate depression and applied externally to heal burns, cuts, and abrasions. Its yellow flowers contain the active ingredient hypericin, which has antidepressant, antiinflammatory, and antibacterial properties. Flavonoids have also been identified in the flowers, and this finding leads researchers to suggest St. John's wort also may boost the immune system.

Uses of St. John's Wort: St. John's wort is used to treat depression (page 215) and fibromyalgia (page 236).

Dosage Information: Look for a standardized hypericin content of 0.3 percent in capsules, tincture, and extract. It is also available as dried leaves and flowers (containing 0.2 to 2.0 percent hypericin) to make infusion and as an ointment for topical use. St. John's wort should be taken with meals to avoid gastric upset. To prepare an infusion from the dried leaves and flowers, add 1 to 2 teaspoons dried herb to 8 ounces boiling water. Cover and allow to steep for fifteen minutes. Drink up to 3 cups daily. Follow dosage recommendations and application directions on commercial ointments and other products.

Possible Side Effects: St. John's wort should be taken under a doctor's supervision. In a few people this herb can cause high blood pressure, headache, stiff neck, nausea, vomiting, and sensitivity to light (photosensitivity).

Possible Interactions: St. John's wort may interact with the amino acids tryptophan and tyrosine; amphetamines; diet pills; nasal decongestants; beverages such as beer, wine, and coffee; foods such as chocolate, fava beans, smoked or pickled foods, yogurt, and salami; and cold or allergy medications. The most common reactions include nausea and high blood pressure.

Tea tree oil

Tea tree oil is derived from the leaves of the *Melaleuca alternifolia tree*, a native of Australia. For centuries, the native Australians have used the oil to treat various skin ailments. Scientific investigations show that the oil contains the active ingredient called terpineol, which is a potent antibacterial and antifungal agent.

Uses of Tea Tree Oil: Tea tree oil is used to treat acne (page 158), athlete's foot (page 186), dandruff (page 214), and eczema (page 231).

Dosage Information: Tea tree oil is an ingredient in some health and beauty products such as toothpaste and soap. Apply the oil directly to the affected area.

Possible Side Effects: If you have sensitive skin, test the oil on a small patch of skin before applying it to a wider area. To avoid getting a skin reaction, dilute the tea tree oil with vegetable oil.

Possible Interactions: None known.

Uva ursi

From the leaves of the uva ursi (*Arctostaphylos uvu-ursi*) shrub comes a substance that has been used to treat urinary-tract problems for at least one thousand years. Uva ursi, also known as bearberry, contains arbutin, which converts to hydroquinone in the urinary tract. It is the hydroquinone that eliminates the infectious agents in the urinary system. It is reportedly especially effective against *E. coli*. Uva ursi can also be applied topically to treat minor skin problems, such as cuts and abrasions, to speed up tissue healing.

Uses of Uva Ursi: Uva ursi is used to treat urinary-tract infections (page 312).

Dosage Information: The recommended forms are capsules and tincture; it is also available as dried leaves and tablets. To make an infusion from the dried leaves, simmer 1 to 2 teaspoons dried leaves in 8 ounces water for five to ten minutes. Drink up to 3 cups daily. Uva ursi requires an alkaline environment in order to be effective; therefore avoid eating acidic foods such as citrus fruit and juices, vitamin C, and sauerkraut during the course of taking this herb. Follow package information for dosage recommendations.

Possible Side Effects: Because uva ursi contains a high level of tannins, it can cause nausea, vomiting, and stomach distress. Do not take this herb for longer than seven days unless you are under the guidance of a medical professional. Pregnant women should not use this herb.

Possible Interactions: None known.

Valerian

An herb known for its calming effect is valerian (*Valeriana officinalis*). Valerian is probably best known for its ability to improve the quality of sleep, hasten onset of sleep, and reduce the number of nighttime awakenings. Unlike conventional sleep aids, valerian does this without side effects, addiction, and withdrawal symptoms associated with barbiturates or benzodiazepines.

Uses of Valerian: Valerian is used to treat anxiety (page 173), headache (page 250), and insomnia (page 279).

Dosage Information: Look for standardized capsules and tinctures; it is also available as dried root for decoctions and tablets. Follow dosage information on product labels.

Possible Side Effects: Do not take valerian if you are also taking conventional tranquilizers or sedatives. Because valerian causes drowsiness, do not drive after taking it. In some individuals, valerian causes excitability, mild headache, or upset stomach. In rare cases it can cause severe headache, nausea, morning grogginess, restlessness, or blurred vision.

Possible Interactions: Valerian should not be used with other medications used to treat anxiety or insomnia.

Wild yam

Wild yam (*Dioscorea villosa*) plants are found in the United States, Mexico, and Asia. The roots of this plant have been found to be a good antispasmodic remedy for the muscles of the intestines and uterus. Wild yam contains steroidal saponins, which account for its soothing properties.

Some confusion surrounds the link between wild yam and progesterone. Wild yam does not contain progesterone, nor does it stimulate the body to produce the hormone. (In the body, progesterone is manufactured from cholesterol.) The confusion stems from the fact that natural pharmaceutical progesterone is made from diosgenin, a constituent of wild yam, which can be transformed into progesterone in a series of chemical reactions in a laboratory. Today, pharmaceutical progesterone is made from yams using a chemical conversion process. Wild yam can be used to make progesterone, but it cannot do so without this pharmaceutical conversion, which cannot be done by the body.

Uses of Wild Yam: Wild yam is used to treat menopausal symptoms (page 287).

Dosage Information: Wild yam is available as a tincture and powdered herb. Follow dosage recommendations on product labels.

Possible Side Effects: Some people experience nausea or vomiting when taking large amounts of wild yam (several times more than the recommended dosage). Do not use during pregnancy or while nursing.

Possible Interactions: None known.

Yohimbe

Yohimbe (*Corynanthe yohimbe*) is derived from the West African yohimbehe tree. It is used as an aphrodisiac among the Bantu tribes in Africa as well as among men in some societies, including those in the

United States. It has also gained much attention as a treatment for impotence. Research has shown that yohimbe contains alkaloids, including yohimbine, which the Food and Drug Administration has recognized as the ingredient that makes yohimbe effective in the treatment of impotence. Yohimbe works by increasing the blood flow to the penis and constricting the veins, which prevents the blood from leaving the penis. This helps maintain an erection.

Uses of Yohimbe: Yohimbe is used to treat impotence (page 270).

Dosage Information: Capsules and tincture are recommended; also available as dried herb and by prescription. Yohimbe should be used under supervision by a physician. Follow package recommendations for dosage information.

Possible Side Effects: Common side effects include nausea, mild hallucinations, chills, dizziness, weakness of the limbs, nervousness, and anxiety. These effects are followed by a relaxed feeling. Yohimbe is an MAO (monoamine oxidase) inhibitor and can cause a severe decrease in blood pressure if it is taken with narcotics, antihistamines, sedatives, alcohol, tranquilizer, or dairy foods. Women should not take yohimbe. Others who should avoid it include anyone with a history of heart or kidney disease, diabetes, high or low blood pressure, duodenal or gastric ulcer, psychiatric illness, and the elderly.

Possible Interactions: Consult your doctor before taking yohimbe if you are taking any prescription drugs. The herb should be avoided by anyone taking blood-pressure-lowering drugs, antidepressants, or antipsychotic drugs.

CHAPTER THREE
Amino Acids and Other
Nutrition Supplements

Many of the products you encounter when browsing the supplement section of the store cannot be classified as vitamins and minerals or herbal remedies. Some of these compounds are amino acids, fatty acids, probiotics, and other natural compounds that assist the body in various metabolic processes. This chapter includes profiles of these products that can be broadly defined as "other" nutrition supplements.

OTHER NUTRITION SUPPLEMENTS A TO Z

Acidophilus

Acidophilus (*Lactobacillus acidophilus*) is a type of beneficial bacteria that resides in the colon and vagina. Acidophilus helps destroy bad bacteria, promotes the growth of good bacteria, improves digestion, and boosts the immune system.

Acidophilus is the main beneficial bacteria in the small intestine, where it replenishes the colonies of good bacteria. Regular supplementation with acidophilus can help maintain a healthy balance of beneficial bacteria and thus inhibit the growth of harmful microorganisms, especially among women who are susceptible to yeast infections. Individuals who are especially likely to be deficient in good bacteria are those taking antibiotics or eating a poor diet, or anyone experiencing diarrhea, especially chronic cases. Active live cultures of acidophilus are found in some brands of yogurt and in acidophilus milk, but the concentration of acidophilus is higher in the supplement.

Uses of Acidophilus: Acidophilus is used to treat athlete's foot (page 186), canker sores (page 200), chronic fatigue syndrome (page 204), constipation (page 211), diarrhea (page 224), diverticulitis (page 226), ear infection (page 228), flatulence (page 237), fibrocystic breast disease (page 234), herpes (page 263), inflammatory bowel disease (page 282), nausea (page 291), and urinary-tract infections (page 312).

Dosage Information: Powder and liquid extracts are available. Acidophilus is available with either a dairy (cow or goat milk) or nondairy (carrot juice, apple pectin) base. The nondairy form is recommended, especially if you are or suspect you may be lactose intolerant. Although acidophilus is also sold as a combination product containing more than one strain of lactobacilli, most experts recommend using a single-strain product. Look for brands that tell you specifically how many CFUs (colony-forming units) you get per dose. Do not buy any product that uses the preservative BHT. Acidophilus is also available in capsule, tablet, and softgel forms. Follow package directions for dosage information. Acidophilus should be taken on an empty stomach and one hour before meals.

Possible Side Effects: No side effects have been reported. People who have an intestinal disorder should consult their physicians before taking acidophilus. Because heat can kill acidophilus, store the supplement in the refrigerator, and do not take it in hot or warm liquids. Be aware of the expiration date on the bottle, because the bacteria must be alive to be beneficial.

Possible Interactions: Use of acidophilus may help promote production of folic acid, biotin, vitamin B6, and niacin.

Arginine

Arginine is an amino acid that helps with wound healing, stimulates the immune system, and assists with the formation of several important hormones, including insulin and growth hormone. It also helps relax the blood vessels, improving blood flow.

Uses of Arginine: Arginine is used to treat cancer (page 193) and infertility (page 277).

Dosage Information: Arginine is available as L-arginine as powder in capsules and tablets; it is also an ingredient in some food bars. It is often used as part of a formula, rather than as a separate supplement. Follow package directions for dosage information.

Possible Side Effects: Side effects are rare at the recommended doses. Nausea, vomiting, flushing, and headache can occur at doses above 20 grams daily.

Arginine can activate the herpes virus in people who already have the virus.

Possible Interactions: Arginine may increase the effect of blood-pressure-lowering drugs.

Brewer's yeast

Yeast is a single-celled organism that multiplies rapidly and is rich in many nutrients, including sixteen amino acids, all the B vitamins except B12 (some forms do contain B12), at least fourteen minerals, and seventeen vitamins. Brewer's yeast can be grown from several different sources: from hops, when it's known as nutritional yeast; or from blackstrap molasses or wood pulp, which produces tourla yeast. Brewer's yeast differs from baker's yeast in that the latter has live yeast cells that deplete the body of B vitamins and other nutrients. These live cells are eliminated in brewer's yeast.

Uses of Brewer's Yeast: Brewer's yeast is used to treat diarrhea (page 224).

Dosage Information: Look for high-quality brands that contain about 60 micrograms chromium per tablespoon or tablet. (Only true brewer's yeast has chromium.) Flakes and powder can be added to food or beverages. Follow dosage information on product labels.

Possible Side Effects: None reported.

Possible Interactions: None known.

Bromelain

Bromelain is a mixture of protein-digesting enzymes found in pineapple. It inhibits the action of proinflammatory substances and prevents the production of compounds produced during the inflammatory process that cause pain. Research shows that it also promotes circulation, reduces the stickiness of platelets in the blood, reduces bruising, and aids in digestion.

Uses of Bromelain: Bromelain is used to treat bursitis (page 191), carpal tunnel syndrome (page 200), and varicose veins (page 314).

Dosage Information: Tablets and capsules are preferred. The dosage of bromelain is based on its milk clotting unit (mcu) activity or on gelatin dissolving units (gdu). For dosage information, follow package directions. Take on an empty stomach when used as an antiinflammatory and with meals when needed as a digestive aid.

Possible Side Effects: Bromelain is considered to be very safe. Nausea, vomiting, diarrhea, and heavy menstrual bleeding have been reported on very rare occasions. Do not take bromelain if you have gastric ulcers.

Possible Interactions: If you take bromelain and hydrochloric acid, the acid will destroy the enzymes.

Carnitine

Carnitine (L-carnitine) is a vitaminlike substance produced in the body from two other amino acids, lysine and methionine, with the help of several other nutrients. L-carnitine stimulates the breakdown of fat to produce energy and prevents the buildup of fat in the heart, skeletal muscles, and liver. It is also used to help infertility in men and to improve muscle strength in people who have neuromuscular conditions. It's been shown that people with some types of muscular dystrophy need greater amounts of carnitine.

Carnitine deficiency is associated with diabetes, cirrhosis, or heart conditions in which oxygen deprivation occurs. Insufficient oxygen supply prevents the heart from storing adequate amounts of carnitine, which leads to an increased risk for angina and heart disease. Meat and dairy products contain the highest amounts of carnitine; yet people who eat little or none of these foods will not experience carnitine deficiency if they get sufficient amounts of lysine, which is found in lysine-fortified grains and in vegetables and legumes, and if they also consume adequate levels of vitamins B1 and B6, and iron.

Dosage Information: Look for free-form L-carnitine in capsules or tablets. Do not get carnitine in any of its other forms: D-carnitine, DL-carnitine, or acetyl-L-carnitine. (The L and D denote a characteristic of the chemical structure.) For dosage information, follow package information. Take carnitine thirty minutes before or after meals.

Possible Side Effects: None when taken as directed. At doses 6,000 milligrams per day or higher, neurological damage may occur. Carnitine should be taken on an empty stomach, as amino acids compete with each other for absorption. Do not take carnitine for more than two months at a time.

Possible Interactions: Carnitine boosts the effectiveness of vitamins C and E.

Chondroitin sulfate

Chondroitin sulfate is a key component of joint cartilage; it is the substance responsible for the gel-like, shock-absorbing lubricant in the joints. Studies indicate that chondroitin sulfate stimulates the synthesis of cartilage, making it useful in the treatment of osteoarthritis. It also helps reduce pain and improves joint function among people suffering from joint disease. Typically it takes three or four months to see results, and six to twelve months to reach maximum benefit.

Uses of Chondroitin Sulfate: Chondroitin sulfate is used to treat osteoarthritis (page 176).

Possible Side Effects: None when taken at recommended dosages.

Possible Interactions: None known.

Coenzyme Q10

The other name for this vitaminlike substance is ubiquinone, a term that reflects the fact that it is found in nearly every cell in the body. Coenzyme Q10 is most concentrated in the mitochondria of cells, where energy is produced, and is most abundant in the heart. Energy production is this enzyme's main role: it helps transform food into ATP, or adenosine triphosphate, the energy that makes the body function. In fact, coenzyme Q10 has a role in 95 percent of the energy generated by the body. It is also a potent antioxidant that has the ability to interfere with histamine, a substance that causes symptoms of asthma and other respiratory conditions.

Much is still unknown about coenzyme Q10. Deficiency appears to be caused by problems with synthesizing the nutrient rather than low intake from diet. Low levels of coenzyme Q10 are often seen in people who have heart problems, AIDS, gingivitis, and in the elderly. Coenzyme Q10 is used to treat people in all these populations as well as those with yeast infections, allergies, cancer, diabetes, high blood pressure and liver problems, as well as obesity, Alzheimer's disease, and muscular dystrophy.

Uses of Coenzyme Q10: Coenzyme Q10 is used to treat allergies and asthma (page 160), Alzheimer's disease (page 164), angina (page 171), cancer (page 193), chronic fatigue syndrome (page 204), diabetes (page 219), gingivitis (page 242), heart attack and cardiovascular disease (page 255), hypertension (page 266), and obesity (page 293).

Dosage Information: Softgel capsules are preferred over capsules and tablets because they are assimilated better. Many softgels also contain vitamin E. The softgels should be bright yellow to orange; these colors indicate that the formulation is pure. Coenzyme Q10 is fairly expensive. Be aware that there are some inferior "bargain" brands on the market, which lure you with reduced prices but which also contain fillers—wheat, corn meal, yeast, eggs, sugars, and/or milk derivatives—in place of the enzyme.

There is no DRI for coenzyme Q10. Recommended dosage levels range from 30 to 150 milligrams per day, although individuals with specific health problems usually take higher doses. To optimize the benefits and absorption of coenzyme Q10, take the supplements with a little fat, such as peanut butter or oil, or take a coenzyme Q10 supplement that contains vitamin E.

Possible Side Effects: Individuals with congestive heart failure who take coenzyme Q10 should not stop taking the supplement without consulting their physician. Do not take this supplement if you are pregnant or breast-feeding. Because coenzyme Q10 is perishable, keep it away from light and heat; it deteriorates in temperatures greater than 115°F.

Possible Interactions: Many drugs—including cholesterol-lowering drugs, beta-blockers, antidepressants, and psychiatric drugs—lower coenzyme Q10 levels in the body.

DHEA

Dehydroephiandrosterone (DHEA) is a hormone produced by the adrenal glands, which is eventually converted into other hormones in the body, such as progesterone, testosterone, and estrogen. Because DHEA is a naturally occurring substance, it cannot be patented, which discourages drug companies from spending money on research.

Because DHEA levels are highest in persons who are younger than forty and decline rapidly thereafter, many people believe it is an antiaging agent. This claim has not been proven or disproven, yet the supplement has gained much popularity for this purpose, as well as for people with Alzheimer's disease.

Uses of DHEA: DHEA is used to treat Alzheimer's disease (page 164) and obesity (page 293).

Dosage Information: Sublingual capsules and liquid are the preferred forms; tablets and capsules are also available. DHEA can be purchased in both nonprescription and prescription strength. Some products sold as "natural" DHEA precursors are derived from wild yam. Despite claims by some manufacturers, wild yam products do not contain DHEA, nor do they transform DHEA in the body. To make this conversion, wild yam must undergo a chemical conversion process in the laboratory. Most products, however, are processed from diosgenin, a substance extracted from wild yam. For dosage information, follow product recommendations.

Possible Side Effects: Minor side effects include acne, mood swings, fatigue, insomnia, enlarged breasts in men, and unwanted body hair (facial hair) in women. Some experts believe liver damage is possible. Because no long-term (more than six months) studies have been done with DHEA, the benefits or disadvantages of prolonged use are unknown.

Precautions: Take DHEA only under medical supervision. People with a personal or family history of any hormone-related cancer, such as prostate or breast cancer, should not take DHEA, because DHEA is a precursor of estrogen and testosterone. Some experts believe that taking high levels of DHEA interferes with the body's ability to naturally synthesize the hormone.

Possible Interactions: If you are taking estrogen, DHEA may affect your dosage requirements. DHEA also may interact with over-the-counter or prescription drugs, nutritional supplements, or herbal remedies.

Fiber

Fiber is divided into two categories—water soluble and water insoluble. Insoluble fiber is a good laxative, but it does not have some of the other benefits of soluble fiber, such as its ability to lower cholesterol and triglyceride levels.

Uses of Fiber: Fiber is used to treat ulcers (page 309) and varicose veins (page 314).

Dosage Information: Fiber supplements are available as capsules, tablets, and powders. Powder forms, mixed with water or juice, are usually the easiest to take. Follow dosage information on product labels. Drink at least 8 ounces of water each time you take fiber supplements. When taking fiber or any bulk laxative, start out with a small dosage and gradually increase the dosage to minimize risk of side effects.

Possible Side Effects: Abdominal gas, sometimes resulting in flatulence and cramping, is possible since soluble fiber is fermented in the intestinal tract. In addition, some people experience allergic reactions to psyllium and other fiber sources.

Possible Interactions: The fiber can decrease the absorption of any medication. Take all supplements and prescription medications at least three hours before or after using fiber supplements.

Flaxseed oil

Flaxseed oil contains two essential fatty acids—alpha-linolenic acid (an omega-3 fatty acid) and linoleic acid (an omega-6 fatty acid). It contains 58 percent alpha-linolenic acid, much more than other vegetable oils. The body often converts ALA into eicosapentaenoic acid (EPA), an omega-3 oil found in fish oils. These essential fatty acids have been found to have cancer-fighting properties (especially against breast cancer), in addition to helping to combat cardio-vascular disease.

Uses of Flaxseed Oil: Flaxseed oil is used to treat psoriasis (page 306).

Dosage Information: The typical dosage is 1 table-spoon daily. Mixing flaxseed oil with other foods, such as yogurt or cottage cheese, can help break up the oil, aiding digestion of the fatty acids.

Possible Side Effects: None known.

Possible Interactions: None known.

Glucosamine (sulfate)

Glucosamine is a substance found naturally in the body. Its primary function is to provide the joints with the material necessary to produce glycosaminoglycan, a critical ingredient in cartilage. It also helps in the formation of tendons, skin, bones, nails, and ligaments. Glucosamine sulfate is the artificially produced version of the natural substance. Studies indicate that glucosamine relieves the pain and stiffness associated with osteoarthritis by stimulating the cells that produce glycosaminoglycans and proteoglycans, another substance important in building cartilage. Glucosamine is also credited with antiinflammatory properties and may be a safe, reliable alternative to aspirin and other antiinflammatory drugs, which often cause dangerous side effects, such as stomach bleeding and stomach ulcers.

Uses of Glucosamine: Glucosamine is used to treat osteoarthritis (page 176).

Dosage Information: Experts have not identified the optimal level of glucosamine in the body. Many arthritis treatments include glucosamine in combination with chondroitin sulfate. Look for brands in which the glucosamine-to-chondroitin dose ratio is 5:4 (for example, 500 milligrams of glucosamine to 400 milligrams of chondroitin). Follow dosage information on package labels.

Possible Side Effects: Occasionally individuals experience minor side effects with glucosamine, including heartburn, diarrhea, nausea, indigestion, and stomach upset. Some glucosamine formulations contain sodium chloride (salt), which should be avoided by people with high blood pressure.

Possible Interactions: People taking diuretics may need to take higher levels of glucosamine for optimal benefits.

5-HTP (5-hydroxytryptophan)

The compound 5-hydroxytryptophan, or 5-HTP, is a derivative of the African plant *Griffonia simplicifolia*. This natural supplement is effective in treating conditions in which serotonin levels are low. Once 5-HTP enters the body, it is converted to the neurotransmitter serotonin. In fact, the body is remarkably efficient at this conversion, making 5-HTP as effective as many prescription drugs at raising serotonin levels (with much fewer negative side effects).

Uses of 5-HTP: 5-HTP is used to treat fibromyalgia (page 236), insomnia (page 279), migraine headache (page 250), and obesity (page 293).

Dosage Information: Follow the dosage information on product labels.

Possible Side Effcts: Mild nausea, heartburn, flatulence, and a feeling of fullness in the abdomen are possible side effects of 5-HTP.

Possible Interactions: 5-HTP should not be used by people taking antidepressants or drugs to treat migraine headache.

Inositol

Inositol is a B vitamin supplement that the body uses to produce lecithin and to transfer fats from the liver to cells in the body. Inositol also aids in reducing high cholesterol levels. Food sources of inositol include dried beans, chickpeas, lentils, cantaloupe, nuts, citrus, whole-grain products, and wheat germ.

Uses of Inositol: Inositol is used to treat arteriosclerosis (page 183) and diabetes (page 219).

Dosage Information: Inositol is available as inositol monophosphate, inositol hexaphosphate, and inositol hexaniacinate. The typical dosage is 100 milligrams daily; follow package directions for specific dosage recommendations.

Possible Side Effects: None known.

Possible Interactions: None known.

Lecithin

Lecithin consists of a mixture of phospholipids and fatty substances found in soy. Lecithin is the primary ingredient in cell membranes; without it the membranes would harden and the cells would die. Lecithin also is in the protective covering of the brain and muscle and nerve cells. These important functions of lecithin make it a critical nutrient in the prevention of gallstones and heart problems. Its healing qualities have made it the target of studies in the prevention of Alzheimer's disease, Parkinson's disease, and other disorders characterized by involuntary muscle movement. Lecithin also enhances brain function, increases energy levels, and aids in the digestion of fats. Lecithin works by preventing the buildup of cholesterol and other fats in the blood vessels and vital organs and eliminating them from the body.

Dietary sources of lecithin include soybeans, lentils, green beans, chickpeas, cauliflower, corn, eggs, and wheat germ. Lecithin supplements are generally not needed by anyone who eats a well-balanced diet.

Uses of Lecithin: Lecithin is used to treat gallstones (page 239), heart attack and cardiovascular disease (page 255), and Parkinson's disease (page 299).

Dosage Information: When shopping for lecithin supplements, check the ratio of lecithin:phosphatidyl-choline:choline, which should be approximately 50:10:1. For dosage information, follow package directions.

Possible Side Effects: Side effects include nausea, vomiting, bloating, abdominal pain, dizziness, and diarrhea.

Possible Interactions: None known.

Lycopene

Lycopene is a carotene found in tomatoes. It has both antioxidant and anticancer properties (especially against prostate cancer). Studies have found that men who had the largest amount of lycopene in their diets had 21 percent lower risk of prostate cancer compared to those who consumed the lowest amount of lycopene.

Similar results have been found with cancer of the gastrointestinal tract, cervix, and breast.

Uses of Lycopene: Lycopene is used to treat cancer (page 193).

Dosage Information: Lycopene supplements can be derived naturally from tomato extracts or made synthetically in the laboratory. Studies have not shown the natural supplement to be any more beneficial than the synthetic version. Follow dosage information on the product label.

Possible Side Effects: None known.

Possible Interactions: None known.

Lysine

Lysine is an amino acid that has proven effective in the treatment of herpes. Lysine has been shown to have antiviral activity. Studies have found that it is most effective when the dietary intake of arginine (found in chocolate, peanuts, seeds, and nuts) is restricted. In one double-blind study, after six months of use, lysine was rated as effective or very effective in three out of four study participants. Foods high in lysine include most vegetables, legumes, fish, turkey, and chicken.

Uses of Lysine: Lysine is used to treat herpes (page 263).

Dosage Information: Do not take lysine if you have liver or kidney disease. Lysine is available as capsules and tablets. Follow the dosage recommendations on the product label.

Possible Side Effects: None known at recommended dosages.

Possible Interactions: None known.

Melatonin

Melatonin is a natural hormone that controls the body's internal clock. It is produced at night and secreted by the pineal gland, which is located deep within the brain. Levels of melatonin in the body fluctuate within a twenty-four-hour period, with peak production at night and the lowest levels during daylight hours. Aging also has an effect on melatonin levels, as the amount of the hormone in the body decreases with age.

Melatonin levels are typically low in people who have insomnia and in the elderly, which is why these populations have trouble sleeping. People who work night shifts, travel frequently across time zones, or otherwise experience jet lag can benefit from taking melatonin, as it readjusts sleep patterns.

Uses of Melatonin: Melatonin is used to treat insomnia (page 279).

Dosage Information: Look for time-release capsules or tablets. Do not take melatonin during the day. Follow dosage information on package labels. For optimal effectiveness, keep melatonin in the refrigerator.

Possible Side Effects: Some people experience morning grogginess, headache, upset stomach, dizziness, disorientation, and sleepwalking.

Possible Interactions: People who are pregnant or nursing or who suffer with depression, schizophrenia, kidney disease, or an autoimmune disease should not take melatonin. Melatonin should not be used by people who take chemotherapy drugs.

Omega-3 fatty acids and fish oil

Omega-3 fatty acids are perhaps best known for their ability to help prevent heart disease by lowering the blood level of low-density lipoprotein (LDL), which is the harmful type of cholesterol that leads to atherosclerosis. Omega-3 fatty acids are versatile fatty molecules found in certain fish (cold-water fish such as sardines, mackerel, and tuna), walnuts, and canola oil. The world's richest source, however, is flaxseed oil.

There are two types of omega-3's—EPA, eicosapentaenoic acid, and DHA, docosahexaenoic acid. These fatty acids keep the cell membranes pliable so the cells can pass easily through blood vessels and suppress the production of leukotrienes, substances that cause inflammation. Both fish oils and flaxseed oil supply EPA and DHA, although flaxseed oil provides a lesser amount of EPA.

Often inflammation is caused by too much of another omega fatty acid—omega-6, which is found in evening primrose oil and several other plants. Maintaining a proper balance of these two fatty acids is important to help prevent pain related to inflammation. Flaxseed oil contains a balance of these two fatty acids.

Uses of Omega-3 Fatty Acids: Omega-3 fatty acids are used to treat irritable bowel syndrome (page 282), psoriasis (page 306), and rheumatoid arthritis (page 180).

Dosage Information: Flaxseed oil is the preferred source of omega-3 fatty acid and contains more than twice the amount of omega-3 found in fish oil. Look for organic, unrefined flaxseed oil, available either as oil or in capsules. Flaxseed oil can be used in salads or on vegetables. Do not use it for cooking. Follow package directions for dosage information.

Possible Side Effects: Fish oil can cause blood sugar levels and LDL cholesterol levels to rise. Nosebleeds can occur because fish oil thins the blood, and gastrointestinal problems, along with burping "fishy" smells, affect some people. People with heart disease or diabetes should consult with their physicians before taking fish oil. Flaxseed oil does not cause side effects.

Possible Interactions: Fish oil can increase the anticoagulant effects of anticoagulant medications, including aspirin. Consult with your physician before starting any omega-3 supplementation.

Quercetin

Quercetin belongs to a group of nutrients known as bioflavonoids (or flavonoids), which are water-soluble plant pigments. It is a powerful antioxidant, antihistamine, and antiinflammatory agent. The best food sources of quercetin are apples, black tea, citrus fruit, and onions, with lesser amounts found in leafy green vegetables and beans.

Uses of Quercetin: Quercetin is used to treat allergies and asthma (page 160), arteriosclerosis (page 183), eczema (page 231), and gingivitis (page 242).

Dosage Information: Quercetin is available as a single nutrient or in combination with other bioflavonoids. It is available as powder, tablets, and capsules; for dosage information, follow package recommendations.

Possible Side Effects: Quercetin can cause diarrhea if taken in extremely high doses (more than 5,000 milligrams daily).

Possible Interactions: None known.

SAMe (S-adenosylmethionine)

SAMe—officially known as S-adenosylmethionine—is involved in more than forty biochemical reactions in the body, many involving folic acid and vitamin B12 as well. SAMe is required for the manufacture of sulfur-containing compounds in the body. Because SAMe is used in so many metabolic processes, it can be useful in the treatment of many disorders and diseases. It has a positive track record as an antidepressant, as well as a treatment of fibromyalgia and migraine headache.

Uses of SAMe: SAMe is used to treat depression (page 215), fibromyalgia (page 236), and migraine headaches (page 250).

Dosage Information: SAMe is available as capsules and tablets. Follow dosage information on the product label. The typical dosage is 200 to 400 milligrams twice daily.

Possible Side Effects: Nausea and gastrointestinal disturbances are possible, but infrequent.

Possible Interactions: None known. People with manic depression should not take SAMe without a doctor's recommendation and supervision.

Soy

Soy isoflavone compounds are considered phytoestrogens, meaning they are naturally occurring plant compounds that bind to estrogen receptor sites in humans. They have been found to reduce hot flashes and other complaints associated with menopause.

Uses of Soy: Soy is used to treat menopausal symptoms (page 287).

Dosage Information: Soy isoflavones are available as capsules, tablets, and powder. The typical dose to relieve menopausal symptoms is about 100 milligrams daily; follow product recommendations for specific dosage information. (One cup of soybeans provides about 300 milligrams of isoflavones.)

Possible Side Effects: Because soy isoflavones work by binding to estrogen and progesterone receptors, some experts fear that they may encourage the growth of breast and uterine cancers. Although studies to date have not confirmed this fear, if you have a history of reproductive cancers, you should avoid soy isoflavones.

Possible Interactions: Women taking birth control pills should avoid using soy isoflavones.

Sulfur

Sulfur is a mineral used by the body to manufacture the bile it needs for digestion. It is found in all the body's tissues, especially those with a high percentage of protein. Sulfur also is a major component of a substance called keratin, which is found in skin, hair, and nails. (If you have ever smelled burning hair, the odor you smell is sulfur.) The body uses sulfur to build keratin and collagen, produce insulin, move nutrients and waste materials in and out of cells, and minimize inflammation and pain.

The bioavailable form of sulfur is methylsulfonylmethane (MSM), which is found in legumes, whole grains, vegetables, meat, and dairy products. Because MSM is easily destroyed during food processing and cooking, the most reliable food sources are unprocessed, uncooked foods.

Dosage Information: Scientists have not yet determined a recommended or suggested dosage for sulfur.

Uses of Sulfur: Sulfur is used to treat bursitis (page 191) and heartburn (page 253).

Dosage Information: Sulfur is available as capsules and powder. The powder can be dissolved in any nonalcoholic liquid or mixed with food. Creams are available for topical use. Follow dosage information on product labels. Take sulfur with food to avoid possible gastrointestinal problems.

Possible Side Effects: Some people experience headache and gastrointestinal problems when taking sulfur at the 3,000-milligram level.

Possible Interactions: Sulfur may interact with blood-thinning drugs such as aspirin or heparin. If you are taking blood thinner, consult your physician before taking MSM.

CHAPTER FOUR
Homeopathic Remedies

In the past few years, homeopathic remedies have joined herbs and vitamins on supplement shelves. While most Americans remain unfamiliar with the theories behind homeopathy, many have found healing and health by simply following the package directions.

Homeopathy is based on the premise that less is more. This holistic approach to healing relies on the use of infinitesimal amounts of substances—plants, minerals, chemicals, microorganisms, animal materials, and even modern drugs—to boost the body's defenses against illness.

To choose an active ingredient, homeopaths turn to the law of similars—or "like cures like"—one of the tenets of homeopathy, which states that a remedy taken in small amounts will cure the same symptoms that it would cause if taken in large amounts. In fact, the world homeopathy has its roots in the Greek words *homo*, meaning "like, similar" and *pathos*, meaning "suffering or disease."

As an example, consider the medicinal use of belladonna, an extract from the poisonous plant deadly nightshade. If taken internally, belladonna can cause high fever, flushed face, confusion, and other flulike symptoms; when used as a homeopathic remedy, belladonna is used to treat fever and flu. Of course, the actual amount of belladonna used in the homeopathic remedy is dramatically diluted. In fact, the active ingredients in homeopathic remedies are diluted to such a degree that not a single molecule of the active ingredient can be found in the solution in its final form. Despite the dilution, evidence indicates that homeopathic remedies work, as you will see later in this chapter.

The dependence on dilution illustrates the second law of homeopathy, the law of infinitesimals. This theory states that the smaller the dose of an active ingredient, the more effective the cure. However, for the less-is-more principle to work, each time the solution is diluted, it must be "potentized" (or shaken) to create "memory of the energy" that cures the body. The law of infinitesimals was discovered through careful experimentation by Dr. Samuel Christian Hahnemann (1755–1843), a German physician who became the founder of the practice of homeopathy.

A BRIEF HISTORY OF HOMEOPATHY

In the early nineteenth century, Dr. Hahnemann grew frustrated by the crude and often dangerous medical practices of the day, including bloodletting and the use of massive doses of drugs that caused diarrhea and vomiting. Not surprisingly, these harsh measures often caused patients more harm than good.

Hahnemann dismissed the traditional practice of medicine and set forth many radical ideas, including the importance of a balanced diet, regular exercise, fresh air, and less confined housing. His ideas were not well received, and Hahnemann eventually gave up the practice of medicine and became a translator. In 1790, while translating *A Treatise on Materia Medica* by Dr. William Cullen, Hahnemann encountered a passage about Peruvian bark (cinchona). Cullen had written that quinine, which was purified from the bark of the cinchona tree, could be used to treat malaria because of its astringent qualities. Hahnemann, a chemist by training, was baffled by this statement because he knew there were many other astringent substances that had no effect whatsoever on malaria. In response, he began to experiment.

First, Hahnemann took quinine himself, only to find that he developed the symptoms of malaria, including fever, chills, thirst, and a pounding headache. Through his observations during a series of experiments—both on himself and on his friends—he developed the homeopathic law of similars. He concluded that the quinine triggered the body's defenses against malaria because the drug caused the symptoms similar to those caused by the disease itself.

In further experiments, known as "provings," Hahnemann observed that the substances caused significant side effects when used in high concentrations, but he could dilute the medication and protect the healing powers through "potentization." In fact, his tests led Hahnemann to conclude that by repeatedly diluting a substance with distilled water or alcohol and shaking it vigorously between dilutions, he could actually increase the strength or potency of the medicine.

In the nineteenth century, Hahnemann's theories were put to the test in dealing with epidemic disease, such as cholera, typhoid, and yellow fever. Homeopathic remedies were credited with saving many lives, and interest in homeopathy spread across Europe. (Modern critics claim that many conventional nineteenth-century treatments actually increased fluid loss, exacerbating illness and causing dehydration and death. Homeopathic remedies, on the other hand, allowed the illness to run its course without causing further head problems.)

In the United States, the first homeopathic college opened in Philadelphia in 1836, and eight years later a group of homeopaths formed the American Institute of Homeopathy, the first national medical organization in the country. Two years later, the American Medical Association was formed, in part to stem the tide of defections by conventional physicians to homeopathy. By the end of the century, there were 15,000 homeopaths and twenty-two schools of homeopathy nationwide.

Homeopathy also flourished and continues to thrive in Europe, particularly in Great Britain, where the Queen of England has her own homeopathic physician and the British National Health Service covers the cost of homeopathic treatments. Today more than 40 percent of British, French, and Dutch doctors use homeopathy or refer some patients to homeopaths, as do 20 percent of German physicians.

At the end of the nineteenth century, fully one out of every five American doctors practiced homeopathy, but by the middle of the twentieth century the American practice had all but died out. One reason was that professional medical groups often expelled physicians who practiced homeopathy or consulted with homeopaths; another factor was that the discovery of antibiotics and other advances in modern medicine lured people to support a more "scientific" approach to healing.

Today homeopathy is experiencing a comeback. Until recently Americans put their trust in high-tech healing, but in the last decade or so they have grown frustrated with the limitations of conventional medicine and turned back to low-tech, natural treatments for many medical problems. As a result, the two-hundred-year-old practice of homeopathy has once again become a popular option.

PUTTING HOMEOPATHY TO THE TEST

Even the staunchest supporters of homeopathy must admit that its theories defy many of the laws of physical science as we know them. Skeptics charge that any healing that takes place stems from the placebo effect, the benefits a patient gets just by believing that a particular treatment will work. However, a number of studies published in respected medical journals have shown that homeopathic remedies work. Consider the evidence.

- 1986: Researchers in Scotland gave a homeopathic hay fever remedy to 144 people with a pollen allergy. To eliminate prejudice on the part of the researchers, the study was double-blind and placebo-controlled, meaning that neither the researchers nor the participants knew who was receiving the homeopathic remedy and who was receiving the placebo. Compared with those who took the placebo, the homeopathic group showed a significant reduction in symptoms, and their need for antihistamines dropped by 50 percent.

- 1988: A French allergist (who was not a homeopath) diluted an antibody 120 times, to the point that the final solution contained no measurable amount of the original antibody. But the final solution still had a noticeable effect on white blood cells, just as diluted homeopathic remedies have an effect on illness.

- 1989: Nonhomeopathic researchers in Great Britain conducted a double-blind, placebo-controlled study to test a homeopathic flu remedy on nearly 500 people. After two days the homeopathic remedy had relieved twice as many flu symptoms as the placebo.

- 1990: German researchers tested eight homeopathic remedies on sixty-one people with varicose veins as part of a double-blind, placebo-controlled study. Among the participants who took the homeopathic remedies, symptoms improved by 44 percent; among those who took the placebo, symptoms became 18 percent worse.

- 1991: A British medical journal published an analysis of 105 clinical studies involving the efficacy of homeopathy. The homeopathic treatment was found to be more effective than a placebo in eighty-one of the studies. Critics of homeopathy charged that many of the studies were poorly designed, but a review of twenty-six of the better-controlled studies found that fifteen demonstrated the benefit of homeopathic treatments.

- 1994: A study published in the British medical journal *The Lancet* found that homeopathic treatment outperformed a placebo in bringing relief to twenty-eight patients allergic to dust mites.

- 1994: The peer-reviewed American medical journal *Pediatrics* reported that among eighty-one children in Nicaragua treated for diarrhea, those given a homeopathic treatment in addition to the standard oral rehydration therapy got well faster than those who got the standard treatment alone. Among the children in the control group, the diarrhea lasted an average of four days, but in the group receiving the homeopathic treatment, it lasted two and a half days.

Despite the mounting evidence, no one really understands exactly how or why homeopathy works. Some researchers have speculated that the "potentization"—the repeated diluting and shaking of the substance—created a distinctive electrochemical pattern in the water. According to this theory of energy medicine, when a patient takes the homeopathic remedy, the electrochemical pattern in the solution somehow subtly changes the electromagnetic fields in the body.

Not all scientists accept this explanation, and ultimately the questions of how homeopathy works remain open. But you can take advantage of the healing benefits of homeopathy without understanding all of its mysteries. All you need to do is keep an open mind and learn more about how you can use homeopathy to manage a wide range of common ailments and medical complaints.

TREAT THE PERSON, NOT THE DISEASE

Homeopaths and traditional doctors approach healing from different points of view. Homeopaths believe illness is not localized in one organ or manifested in one symptom, so when prescribing treatment they consider the entire person, both the mind and body. Practitioners of conventional medicine, on the other hand, tend to focus on suppressing symptoms, taking little or no account of the person's emotional or overall physical condition. Homeopaths consider physical symptoms as positive signs that the body is hard at work defending and healing itself. Rather than trying to eliminate symptoms, homeopathic remedies sometimes even aggravate symptoms for a time as they stimulate the body's self-healing mechanism.

Because homeopaths prescribe treatments based on a variety of very specific symptoms reported by the patient, the home use of homeopathic remedies should be limited to treatment of relatively minor ailments, such as colds, runny noses, and other non-life-threatening conditions. If you experience recurrent or potentially dangerous symptoms, consider consulting a trained homeopath or other health care professional.

You can find a wide range of homeopathic remedies at natural food stores and pharmacies. They are labeled for their intended use and can be used like other over-the-counter medicines. Because the active ingredients have been diluted, homeopathic remedies do not cause unwanted side effects.

PRACTICING HOMEOPATHY

Homeopathic remedies are prepared according to standards of the United States Homeopathy Pharmacopoeia and come in a variety of potencies, based on the strength of dilution. The three most common forms of remedies are the mother tincture, x potencies, and c potencies.

- *The mother tincture.* The mother tincture is an alcohol-based extract of a specific substance, such as the plant yellow jasmine or the mineral potassium carbonate. They are usually used topically.

- *X potencies*. The x represents the number ten. In homeopathic remedies with x potencies the mother tincture has been diluted to 1 part in 10 (1 drop of tincture to every 9 drops of alcohol). The number before the x tells how many times the mother tincture has been diluted. For example, a 12x potency represents 12 dilutions of 1 in 10. The more the substance is diluted, the more potent it becomes, so a remedy with a 30x potency is considered stronger or more potent that one with a 12x potency.

- *C potencies*. The c represents the number 100, so homeopathic remedies with a c potency have been diluted to 1 part in 100 (1 drop of tincture to every 99 drops of alcohol), making them much stronger than x potencies. Again, the number before the c represents the number of dilutions. A 3c potency represents a substance that has been diluted to one part in 100 three times; by the time 3c is reached, the dilution is 1 part per million. In general, 6c is the potency recommended for most acute or self-limited ailments, and 30c for chronic conditions or emergencies.

When you use the right remedy, it will work quickly and you can discontinue treatment. The wrong remedy will cause no harm, but you will not improve. Because the amount of active ingredient is so small, side effects from homeopathic remedies are virtually nonexistent, even among infants and the elderly.

TREATMENT TIPS

- Homeopathic remedies come in pellet, tablet, and liquid forms. Avoid touching them; instead, shake the pellets into a measuring spoon and place them under your tongue and allow them to dissolve.

- Homeopathic tinctures (liquid remedies) contain alcohol. Homeopathic creams, ointments, and salves can be made by mixing diluted remedies with cream or gel base for topical use.

- Take a homeopathic remedy at least a half hour before or after eating. Strong flavor can decrease the effectiveness of remedies since the impact of the homeopathic remedy is very subtle.

- Odors can also affect efficacy; if possible, avoid strong smells, such as perfumes, chemical odors, and other scents for a half hour before or after taking a homeopathic remedy.

- If you take a treatment for a condition that clears up in fifteen minutes, you don't need to take another dose. If the symptoms later return, you can take a second dose. Should the second attempt also fail, contact a homeopath or your primary doctor, as needed.

COMMON HOMEOPATHIC REMEDIES A TO Z

The following homeopathic remedies are used to treat common medical conditions described in this book. The list includes both the common abbreviations and the full Latin name for each of the remedies, to avoid confusion when purchasing products.

Aconite *(Aconitum napellus)*

- Made from wolfsbane, blue aconite.

- Used to treat bronchitis (page 188), cardiovascular disease (page 255), colds (page 207), insomnia (page 279), and rheumatoid arthritis (page 180).

Agnus *(Agnus castus)*

- Made from chaste tree.

- Used to treat impotence (page 270).

Allium *(Allium cepa)*

- Made from red onion.

- Used to treat bronchitis (page 188) and colds (page 207).

Alumina *(Alumina)*

- Made from aluminum oxide.

- Used to treat Alzheimer's disease (page 164) and constipation (page 211).

Amyl *(Amyl nitrite)*

- Made from amyl nitrite.

- Used to treat menopausal symptoms (page 287).

Arnica *(Arnica montana)*

- Made from Leopard's bane, Fallkraut.

- Used to treat arthritis (pages 176, 180), cardiovascular disease (page 255), gout (page 247), and osteoporosis (page 296).

Arsenicum *(Arsenicum album)*

- Made from arsenic trioxide.

- Used to treat anxiety (page 173), asthma (page 160), and flu (page 207).

Aurum *(Aurum metallicum)*

- Made from gold.

- Used to treat depression (page 215).

Baryta Carb. *(Baryta carbonica)*

- Made from barium carbonate.

- Used to treat hypertension (page 266).

Berberis *(Berberis vulgaris)*

- Made from common bayberry.

- Used to treat gallstones (page 239).

Bryonia (Bryonia alba)

- Made from white bryony.

- Used to treat bronchitis (page 188), constipation (page 211), osteoarthritis (page 176), and menopausal symptoms (page 287).

Calcarea (Calcarea carbonica)

- Made from calcium carbonate.

- Used to treat anxiety (page 173), cataracts (page 202), diarrhea (page 224), menopausal symptoms (page 287), and obesity (page 293).

Calcarea Phos. (Calcarea phosphorica)

- Made from calcium phosphate.

- Used to treat osteoarthritis (page 176).

Causticum (Causticuum hahnemanni)

- Made from potassium hydrate.

- Used to treat incontinence (page 273).

China (China officinalis)

- Made from Peruvian bark, quinine.

- Used to treat anemia (page 168) and gallstones (page 239).

Coffea *(Coffea cruda)*

- Made from unroasted coffee.
- Used to treat insomnia (page 279).

Colchicum *(Colchicum autumnale)*

- Made from meadow saffron.
- Used to treat gout (page 247).

Conium *(Conium maculatrum)*

- Made from hemlock.
- Used to treat impotence (page 270).

Digitalis *(Digitalis purpurea)*

- Made from red foxglove.
- Used to treat cardiovascular disease (page 255).

Eupatorium *(Eupatorium perfoliatum)*

- Made from boneset.
- Used to treat flu (page 207).

Euphrasia *(Euphrasia officinalis)*

- Made from common eyebright.
- Used to treat bronchitis (page 188) and colds (page 207).

Ferrum *(Ferrum metallicum)*

• Made from iron.

• Used to treat anemia (page 168) and obesity (page 293).

Ferrum Phos. *(Ferrum phosphoicum)*

• Made from iron phosphate.

• Used to treat flu (page 207) and incontinence (page 273).

Gelsemium *(Gelsemium sempervirens)*

• Made from yellow jasmine.

• Used to treat bronchitis (page 188), colds (page 207), and headaches (page 250).

Hamamelis *(Hamamelis virginiana)*

• Made from common witch hazel.

• Used to treat hemorrhoids (page 262) and varicose veins (page 314).

Hyoscyamus *(Hyoscyamus niger)*

• Made from henbane.

• Used to treat Parkinson's disease (page 299).

Ignatia (Ignatia amara)

- Made from St. Ignatius's bean.

- Used to treat anxiety (page 173), depression (page 215), impotence (page 270), insomnia (page 279), and nausea (page 291).

Kali Carb. (Kali carbonicum)

- Made from potassium carbonate.

- Used to treat asthma (page 160) and obesity (page 293).

Lachesis (Trigonocephalus lachesis)

- Made from surucucu or bushmaster snake venom.

- Used to treat cardiovascular disease (page 255) and menopausal symptoms (page 287).

Ledum (Ledum palustre)

- Made from wild rosemary.

- Used to treat gout (page 247) and osteoarthritis (page 176).

Lycopodium (Lycopodium clavatum)

- Made from club moss.

- Used to treat Alzheimer's disease (page 164), anxiety (page 173), constipation (page 211), and impotence (page 270).

Nux *(Strychnos nux vomica)*

• Made from poison nut tree.

• Used to treat cardiovascular disease (page 255), constipation (page 211), depression (page 215), flu (page 207), gingivitis (page 242), incontinence (page 273), and insomnia (page 279).

Phosphoric acid *(Phosphoricum acidum)*

• Made from phosphoric acid.

• Used to treat diabetes (page 219).

Phosphorus *(Phosphorus)*

• Made from phosphorus.

• Used to treat Alzheimer's disease (page 164), anxiety (page 173), and bronchitis (page 188).

Pulsatilla *(Pulsatilla nigricans)*

• Made from wind flower.

• Used to treat bronchitis (page 188), depression (page 215), flu (page 207), incontinence (page 273), menopausal symptoms (page 287), rheumatoid arthritis (pages 176, 180), and varicose veins (page 314).

Rhododendron *(Rhododendron chrysanthum)*

• Made from Siberian rhododendron.

• Used to treat rheumatoid arthritis (page 180).

Rhus Tox. *(Rhus toxicodendron)*

- Made from poison ivy.
- Used to treat cardiovascular disease (page 255), eczema (page 231), and osteoarthritis (page 176).

Sanguinaria *(Sanguinaria canadensis)*

- Made from bloodroot.
- Used to treat bronchitis (page 188).

Sepia *(Sepia officinalis)*

- Made from cuttlefish ink.
- Used to treat menopausal symptoms (page 287).

Silicea *(Silicea terra)*

- Made from flint.
- Used to treat acne (page 158), cataracts (page 202), constipation (page 211), and diabetes (page 219).

Symphytum *(Symphytum officinale)*

- Made from common comfrey.
- Used to treat osteoporosis (page 296).

Urtica *(Urtica wrens)*

- Made from small nettle.
- Used to treat gout (page 247).

HOW TO FIND A QUALIFIED HOMEOPATH

Homeopaths practice a form of medicine, and homeopaths are considered medical professionals. The requirements for licensure vary from state to state. While medical doctors (M.D.s) and osteopaths (D.O.s) legally can practice homeopathy in all states, many other medical professionals—such as naturopaths (N.D.s), and chiropractors (D.C.s)—may be qualified to practice homeopathy, provided they meet the necessary licensing requirements.

For more information about homeopathy or to find a licensed practitioner, contact the following.

NATIONAL CENTER FOR HOMEOPATHY
801 N. Fairfax Street, Suite 306
Alexandria, VA 22314
(703) 548-7790; (877) 624-0613
www.homeopathic.org

INTERNATIONAL FOUNDATION FOR HOMEOPATHY
P.O. Box 7
Edmons, WA 98020
(425) 776-4147
www.healthy.net/library/journals/resonanc/
editors.htm

HOMEOPATHIC EDUCATIONAL SERVICES
2124B Kittredge Street
Berkeley, CA 94704
(510) 649-0294
www.homeopathic.com

≋ CHAPTER FIVE
Creating Your Individualized
Supplement Program

Now that you understand the importance of vitamins, minerals, herbs, and amino acids in general, you must decide which specific products you need to optimize your health. Obviously you don't need to take every supplement described in this book, so the challenge becomes figuring out which ones can best address your particular physical demands.

The following chapter will help you design a supplement program customized to your health profile. Your program will consist of a multivitamin–mineral supplement—the foundation of your program—plus additional supplements, chosen to enhance your health and address your specific health concerns. You can design your supplement program in three steps: Step 1—Choosing a Multivitamin, Step 2—Choosing Additional Core Supplements, and Step 3—Choosing Healing Supplements.

STEP 1: CHOOSING A MULTIVITAMIN

A high-quality multivitamin–mineral supplement is the foundation of your program. Researchers at the

Harvard Medical School announced in the June 2002 issue of the *Journal of the American Medical Association* their recommendation that *all* adults should take a multivitamin in addition to eating a healthy diet to ensure optimal health. Evidence shows that a multivitamin can lower your risk of many diseases, including cardiovascular disease, cancer, and osteoporosis.

Remember, nutritional supplements should supplement a healthy diet, not make up for poor dietary and lifestyle habits. Don't assume you can skip the broccoli and oranges and whole grains and make up for your dietary indiscretions by popping a pill. Good health isn't that easy.

A good multivitamin–mineral supplement should contain all of the essential vitamins and minerals at doses that fall within the ranges listed below. In general, if your dietary habits are good—you eat mostly whole, natural foods, you consume at least five servings of fruits or vegetables daily, you drink pure water, and your fat intake is less than 30 percent—you can go with a supplement with values at the lower end of the ranges. On the other hand, if you subsist on processed foods, you consider French fries a vegetable, and your vision of the ideal breakfast is coffee and a glazed doughnut, you need a high-potency supplement that contains nutrients at the higher end of the scale. (You also need to rethink your diet plan.)

Some multivitamins contain minute amounts of trace minerals as well—for example, silica, vanadium, and iodine. These extra nutrients are not necessary, so do not worry if the supplement you choose does not have these ingredients. When looking for a multivitamin, focus on the key nutrients.

What Should Be in a Multivitamin-Mineral Supplement?

NUTRIENT	RANGE
Vitamins:	
Vitamin A (preferably as beta-carotene)	5,000 IU*
Beta-carotene (if not listed as vitamin A)	5,000–25,000 IU
Vitamin B1 (thiamin)	10–100 milligrams
Vitamin B2 (riboflavin)	10–50 milligrams
Vitamin B3 (niacin)	10–100 milligrams
Vitamin B5 (pantothenic acid)	25–100 milligrams
Vitamin B6 (pyridoxine)	25–100 milligrams
Biotin (Vitamin B7)	100–300 micrograms
Folic acid (Vitamin B9)	400 micrograms**
Vitamin B12	400 micrograms
Vitamin C	100–1,000 milligrams
Vitamin D	100–400 IU
Vitamin E (d-alpha tocopherol)	100–800 IU
Choline	10–100 milligrams
Minerals	
Calcium	250–1,500 milligrams***
Chromium	200–400 micrograms
Copper	1–2 milligrams

* Women of childbearing age should limit their intake of vitamin A to 2,500 IU daily, as there is a risk of birth defects at higher amounts.

** Older people should take a supplement that contains 1 milligram, because higher amounts can hide a severe deficiency of vitamin B12.

*** Best to take as a separate supplement.

Iron	15–30 milligrams*
Magnesium	250–500 milligrams
Manganese	10–15 milligrams
Molybdenum	10–25 micrograms
Potassium	200–500 milligrams
Selenium	100–200 micrograms
Zinc	15–45 milligrams

* Postmenopausal women and men typically do not need additional iron. They should choose a multiple supplement without iron.

Vitamins with nutrients at the higher end of the range are sometimes referred to as "high-potency" vitamins. According to government regulations, "high-potency" means that the product contains 100 percent or more of the Daily Value (DV) for that particular nutrient. If the product contains many nutrients—for example, a multivitamin–mineral supplement—at least 100 percent of the DV is to be classified as "high potency."

More is not always better, and in some cases can be dangerous. According to the *Nutritional Desk Reference*, avoid any supplement that provides more than 300 percent of the DVs for all the ingredients listed. Although it may sound like a good deal, such a supplement can upset the balance of nutrients in the body by providing excessive amounts of some in relation to others.

STEP 2: CHOOSING ADDITIONAL CORE SUPPLEMENTS

Even the most ambitious multivitamin–mineral supplement can't squeeze in all of the calcium, vitamin C, and vitamin E you probably need for optimal health. If a pill contained adequate levels of all of these nutrients, it would be too large to swallow. In addition, you may have other health concerns—say, a heart condition, joint pain, or menopausal symptoms—that can be partially addressed using nutritional supplements. No single pill can meet all your needs, so you should consider taking several other supplements in addition to your multivitamin.

The most common supplements taken in addition to a multivitamin include calcium, vitamin C, and vitamin E. These nutrients prove to be beneficial to most people, both in the prevention and treatment of many health conditions. (For more information on their benefits, see Chapter 1.)

If you want to cut down on the number of pills you take each day, you may want to examine various supplement "formulas," which combine several nutrients into a single supplement. Some manufacturers use formulas as a marketing gimmick, although the approach can be beneficial to people with certain common health problems or conditions who want to swallow fewer pills. One caveat: Review the nutrient profile of various products to make sure that you don't compromise your core supplement in order to make room for extras.

Manufacturers are free to combine nutrients in any way they wish, provided they accurately label their products. No official guidelines dictate the nutrients that must be included in any particular formula. The following list summarizes common ingredients and dosages found in popular formulas.

• *Antioxidant formulas* typically include one or more of the following.

  ❏ Vitamin A (5,000–10,000 milligrams)
  ❏ Vitamin C (1,000–3,000 milligrams)

  ❑ Vitamin E (200–400 IU)
  ❑ Selenium (200–400 micrograms)

- *Energy formulas* typically include one or more of the following.

  ❑ Ginseng (100–200 milligrams)
  ❑ Magnesium (150–300 milligrams)
  ❑ Antioxidants (see above)

- *Heart health formulas* typically include one or more of the following.

  ❑ Carnitine (500–1,500 milligrams)
  ❑ Garlic (5,000 micrograms of allicin)
  ❑ Coenzyme Q-10 (50–100 milligrams)
  ❑ Folic acid (200–400 micrograms)
  ❑ Hawthorn (100–250 milligrams)
  ❑ Antioxidants (see above)

- *Immune support formulas* typically include one or more of the following.

  ❑ Echinacea (150–300 milligrams)
  ❑ Goldenseal (250–500 milligrams)
  ❑ Zinc (15–50 milligrams)
  ❑ Coenzyme Q-10 (50–100 milligrams)
  ❑ Antioxidants (see above)

- *Joint support formulas* typically include one or more of the following.

  ❑ Glucosamine and chondroitin sulfate (500–1,500 milligrams)
  ❑ Boron (1.5–3 milligrams)
  ❑ Antioxidants (see above)

- *Memory support formulas* typically include one or more of the following.

  ❑Ginseng (100–200 milligrams)
  ❑Gingko (40–120 milligrams)
  ❑Coenzyme Q-10 (50–100 milligrams)
  ❑Antioxidants (see above)

- *Menopause support formulas* typically include one or more of the following.

  ❑Black cohosh (250–500 milligrams)
  ❑Evening primrose oil (250–500 milligrams)
  ❑Soy (100–150 milligrams)
  ❑Wild yam (100–300 milligrams)
  ❑Chaste berry (500–1,000 milligrams)

- *Men's formulas* typically include one or more of the following.

  ❑Saw palmetto (100–300 milligrams)
  ❑Lycopene (15–30 milligrams)
  ❑Zinc (15–50 milligrams)
  ❑Antioxidants (see above)

- *Prenatal formulas* typically include one or more of the following.

  ❑Folic acid (200–400 micrograms)
  ❑Calcium (1,000–1,500 milligrams)
  ❑Antioxidants (see above)

- *Weight loss formulas* typically include one or more of the following.

  ❑Chromium (200–400 micrograms)
  ❑Magnesium (200–400 milligrams)
  ❑Carnitine (1,000–3,000 milligrams)

- **Women's formulas** typically include one or more of the following.

  ❑ Calcium (1,000–3,000 milligrams)
  ❑ Iron (5–15 milligrams)
  ❑ Boron (1.5–3 milligrams)

Many of the nutrients found in formulas can be used as an ongoing part of your supplement program. The formula you choose—or the individual supplements you take—may change as you move from one phase of life to another. For example, women of childbearing age face different physical challenges than those passing through menopause, and their supplement programs should change as their physical demands change.

Do You Need One-per-day or Multiple Dosing?

A one-per-day multivitamin-mineral supplement may be convenient, but is it the most effective form to take? Cramming all the high potency you need into one capsule or tablet is hard to do in a size that you can swallow, plus it gives your body a "peak" of nutrients rather than a more even dose throughout the day. Experts generally recommend choosing a supplement that requires you to take two to six capsules or tablets daily. In this way the dose is spread out over the day and your body can assimilate the ingredients better. Also, take the multiple with meals. Multiples taken in between meals can cause stomach distress and are not as well absorbed.

STEP 3: CHOOSING HEALING SUPPLEMENTS

In addition to a multivitamin and additional core supplements that can be taken on an ongoing basis for overall health, you may choose to use supplements to help prevent, treat, and manage other specific health problems. Part 2 of this book, "Prescription for Healing," can help you choose particular products to handle your individual health concerns. Each entry includes a description of a medical problem, followed by a list of supplements that can be useful in treating the condition.

In addition, people with certain health issues or lifestyle factors may benefit from certain supplements. Consider the following:

- People who cannot digest milk products (lactose intolerance) may need to take calcium supplements.

- Strict vegetarians and those who do not eat eggs or dairy products may need to take additional vitamin D, B12, and calcium. In addition, some vegetarians may need to take additional zinc and iron because these nutrients are more easily absorbed from animal products than from plants.

- Homebound individuals and people in nursing homes, regardless of their age, may need a vitamin D supplement because of insufficient exposure to the sun. This situation is of particular concern in high-latitude regions during the winter and spring months.

- People on very-low-calorie diets may need greater amounts of all the vitamins and minerals, but especially vitamin E, calcium, iron, zinc, and vitamin B6, because of their poor nutritional intake.

- Pregnant women need additional amounts of nearly all nutrients during pregnancy. In particular, vitamin B6 is needed because the fetus draws this vitamin from the mother.

- Smokers, people who exercise strenuously, and individuals who are under a great deal of emotional or physical stress usually need additional amounts of antioxidants, especially vitamin C.

- People who drink a great deal of alcohol have difficulty absorbing and utilizing many nutrients, especially the B vitamins. Supplemental B vitamins may improve their nutrient profile.

PULLING IT TOGETHER

Now that you know which products you need, it's time to make a shopping list. Using the information in this chapter and elsewhere in the book, compile a list of the supplements you need. You will want to revisit this list when your health status changes, so that your supplement program always meets your current health demands.

- A multivitamin–mineral supplement: This is the foundation of your supplement plan.

- Additional core supplements: This list includes additional nutrients to enhance your health at dosages not found in your core supplement. (Use the recommendations in this chapter as a guide in making your list.)

☐ _____

☐ _____

☐ _____

- Other healing supplements: This list includes additional supplements needed to address specific health problems or concerns. (Use the recommendations in Part 2 as a guide in making your list.)

    ❑ _____
    ❑ _____
    ❑ _____
    ❑ _____
    ❑ _____

❦ CHAPTER SIX
Shopping for Supplements

Once you decide which vitamins, herbs, and supplements you need, you must sift through the various products crowding the health food and grocery store shelves and determine which particular products you should buy. This is no small task, since you will encounter a number of terms and claims that may be confusing. This chapter will help demystify the information you will find on product labels, making it easier for you to choose the best products at the best price.

Should you choose "natural" vs. "synthetic" vitamins?

The body uses both natural and synthetic vitamins in the same way, so in most cases you don't need to pay more to buy "natural." It would take tons of food to extract all the vitamins used in nutrition supplements sold in the United States. Supplements produced in the laboratory are chemically identical—and much cheaper.

One exception: vitamin E. Natural vitamin E is absorbed better than the synthetic version. Look for products containing natural vitamin E, identified as d-alpha tocopherol.

Should I choose tablets or tinctures, powders or some other form of supplement?

Most supplements can be found in a variety of forms, including tablets, capsules, softgels, tinctures, extracts, powders, and loose herbs. In most cases, your choice is a matter of personal preference. When one form is considered superior to another, it is noted in the profile found in Chapter 1.

One less common form of supplement is liquid spray. Spray supplements can be readily absorbed through the mucous membranes of the mouth, and they don't contain fillers, colors, waxes, and binders used in pills and capsules. On the downside, spray vitamins can be difficult to find, and they tend to be more expensive than other forms of supplements.

What is the % Daily Value (DV) found on supplement labels?

Supplement and food labels list the % Daily Value (DV) of various nutrients, a number indicating how much one serving contributes to the percent of required nutrients in a 2,000-calorie diet. Depending on your age, sex, state of health, and level of physical activity, you may consume more or less than 2,000 calories in your daily diet, meaning you may need more or less than the 100 percent DV.

Despite these recommendations, a significant number of Americans—including those who are obese—fail to consume all the nutrients required for good health. That's not to say these people suffer from an obvious nutritional deficiency, but many do have a marginal deficiency that prevents them from functioning at an optimal level. In fact, government studies have shown that nearly half of all Americans have some kind of nutrient deficiency.

What are Dietary Reference Intakes or DRIs?

In 1997, the Food and Nutrition Board of the National Research Council created the new term Dietary Reference Intake (DRI) to replace the old term Recommended Daily Allowance (RDA). The DRI includes all the nutrients that had been assigned an RDA, plus those that experts believe are important but that have not yet been assigned RDA status. (Nutrients that have not been assigned RDAs are assigned yet another category, Adequate Intakes or AIs.)

The DRIs are based on sex, age, and overall health. (The experts assume that the person is in good health.) The dosages are specifically designed to reduce the risk of chronic disease; they do not take into account factors that can increase a person's requirements for additional nutrients, such as inadequate diet, stress, environmental toxins, drug and alcohol use, smoking, and poor hygiene. For this reason, most physicians and health professionals view DRIs as the absolute *minimum* levels for nutrients and routinely recommend higher dosages for optimal health.

Where should I shop for supplements?

It doesn't matter where you buy supplements, as long as you review product labels carefully. Generally, a health food or vitamin store has a bigger and more diverse selection of nutrition supplements than a pharmacy or grocery store, although many have expanded their inventories in recent years. Health food and vitamin stores are also more likely to carry brands that avoid the use of unnecessary additives. Making purchases through the mail or over the Internet is safe only if you know exactly what you want and have already seen the product in a store and have read the label, or if the label details can be inspected in a catalog or on the Web site. Examples of reputable websites include *www.drweil.com* and *www.lef.org*.

Do I need to worry about additives and fillers?

If you have a food allergy or food sensitivity, you need to watch out for the fillers and additives in nutrition supplements. Look for products that claim to be free of wheat, yeast, milk, salt, soy, cornstarch, and sugar. Typical fillers include talc, rice concentrate, cellulose, silica, and magnesium stearate.

Vegetarians who want to avoid all animal products need to check labels for gelatin, which is made from animals, unless the manufacturer has specified that the capsule is vegetable based. A growing number of supplements use non-animal-based capsules and coatings.

What are "standardized" herbal extracts?

Herbal supplements often bear labels claiming to be a "standardized extract" or "guaranteed potency extract," meaning that the product is guaranteed to contain a standardized or predetermined level of active ingredients. Standardized extracts allow for more accurate dosages.

What are "chelated" minerals?

A chelated mineral, such as chromium picolinate, is one that has been bound to a protein molecule, which carries the mineral to the bloodstream. Some evidence indicates that chelated minerals can be somewhat more easily absorbed (about 5 to 10 percent more for calcium, for example), but chelated minerals cost about five times more than traditional supplements. Check the labels and compare prices, but don't pay much more for chelated minerals.

What does the USP label mean?

The United States Pharmacopeia (or USP) is an independent, nonprofit corporation that sets the standards of quality, purity, strength, packaging, and labeling for drugs and nutritional supplements in the United States. It has been in existence since 1820. Its governing board is composed of more than a hundred representatives from accredited U.S. colleges of medicine and pharmacy, national associations such as the American Medical Association and the National Association of Retail Druggists, and departments of the federal government, including the Food and Drug Administration. Look for "USP" on the label, indicating that the supplement has been tested for dissolvability and overall quality.

How should I store my supplements?

Store your nutrition supplements away from heat and out of reach of children. Keep vitamins A and E in the refrigerator.

How long are supplements good?

While many supplements remain shelf-stable for a year or more, you need to watch the expiration dates printed on product packages. Throw away any vitamins or nutrition supplements that have passed their expiration date; these will not be as effective or potent as they should be.

How should I choose a brand?

Vitamins differ little from one brand to another. In truth, most supplement manufacturers buy their vitamins from the same small group of international suppliers. Brands differ, however, in the amount of each nutrient, in the use of fillers, and in the addition of bonus nutrients, such as ginseng or green tea.

Examine the labels on store-brand products. All vitamins are essentially the same, so forget the brand names and look for the bargain. If you buy a heavily advertised product, all you're doing is paying for the advertising.

How can I find a qualified nutritionist?

If you have special nutritional or health needs, you may want to consult a naturopathic physician or a nutrition counselor for help in designing a supplement regimen. For information on finding a qualified practitioner, see the "Organizations of Interest" section on page 317.

PART 2:

Prescription for Healing

CHAPTER SEVEN
Treating Common Medical Problems

This chapter describes many common medical conditions, diseases, and ailments and lists the most effective supplements you can take to help prevent, treat, or cure them. These supplements should be used in addition to a multivitamin and the additional core supplements you include in your supplement program, as described in Chapter 5.

The entries in this section include a brief description of the medical condition, followed by recommendations of which supplements may prove useful in treating the problem. In most cases, the supplement recommendations are divided into two categories: "Most Helpful Supplements" and "Other Helpful Supplements." As a rule, start by taking one or more of the products listed in the "Most Helpful" section, then adding or deleting supplements from your program based on how you respond to the treatment.

COMMON MEDICAL PROBLEMS A TO Z

Acne

Acne is the most common skin disorder. It starts deep in the pores of the skin, where the sebaceous glands produce sebum, a concoction of oils and waxes designed to keep the skin moist and lubricated. The sebaceous glands—and most cases of acne—are located on the face, back, chest, and shoulders.

Acne usually appears during puberty, when the body produces testosterone and other hormones that increase sebum production. (Both males and females produce testosterone, though males usually generate more of it.) The excess sebum blocks the pores, causing blackheads. When bacteria grow beneath the skin and release enzymes that break down the sebum, the result is whiteheads or pimples.

Acne can appear in several different forms. *Acne vulgaris* is the common form of acne caused by clogged pores and hair follicles. They can develop into a more serious form of acne known as *acne conglobata*, in which cysts or nodules penetrate under the skin and may become inflamed and infected.

Adult-onset acne typically appears in people who are thirty and older; it is often caused by an allergic reaction to cosmetics or food, although some cases are related to menstruation. *Rosacea*, an acnelike condition that mostly affects people over age thirty, begins as a persistent flush on the cheeks and nose that may cause the nose to become thickened, red, and tender, especially in men.

Acne isn't caused by inadequate hygiene or poor diet (too much chocolate, fried foods, and sugar), although these factors can make acne worse. Use of contraceptives and corticosteroids, including anabolic steroids, also can contribute to, but do not cause, acne. Remember, pimples form beneath the surface of the skin, so no amount of surface scrubbing can prevent them entirely.

MOST HELPFUL SUPPLEMENTS

- *Chromium* helps with metabolism of sugar; impaired sugar metabolism may contribute to acne. Studies have found that chromium helps reduce infections of the skin. Take 200 micrograms of chromium picolinate daily with food.

- *Pyridoxine* (vitamin B6) has proven useful for premenopausal acne. Pyridoxine deficiency has been associated with acne. Take 25 to 50 milligrams two to three times daily.

- *Tea tree oil* is considered the ideal skin disinfectant. A solution of 5 percent tea tree oil applied topically, after a thorough cleansing of the area, offers the same benefit as benzoyl peroxide, but without the same drying effects and irritation. Look for commercial products or make your own solution by combining 1 part tea tree oil with 20 parts of olive oil. Commercially prepared tea tree oil soap is another alternative.

- *Zinc* relieves inflammation and heals damaged skin. Zinc inhibits the conversion of testosterone to DHT, the real acne culprit. Take 15 to 25 milligrams of zinc two to three times daily for 3 months. Be patient; it can take weeks or months for the overall condition of your skin to improve.

OTHER HELPFUL SUPPLEMENTS

- *Evening primrose oil* helps heal damaged skin. Take 500 to 1,000 milligrams daily.

- *Vitamin E* reduces scarring and regulates levels of vitamin A in the body. Take 400 IU daily for 3 months.

AN OUNCE OF PREVENTION

Some people tend to be acne prone, while others breeze through life with barely a blemish to show for it. Try to prevent acne by keeping your skin scrupulously clean by washing twice a day with unscented soap. Avoid oily hair products and makeup. Be aware that some chemicals found in acne medication can cause increased sensitivity to the sun.

CALL FOR HELP

If you try natural remedies for three months to no avail, contact a dermatologist for expert help. Also seek medical help if you develop a cyst, pustule, or other skin infection that may require antibiotics.

Allergies and Asthma

It starts out with a simple sneeze. But when one sneeze turns into three, then five, you suspect the worst: allergies. An allergy is the body's response to an irritant you breathe (such as dust mites, pollen, mold, or pet dander), eat (such as dairy products, wheat, or peanuts), or touch (such as wool, fabric softener, or perfume). Inside the body, this perceived invader or allergen is attacked by IgE antibodies (immunoglobulin E). The antibodies trigger the release of chemicals known as histamines, which set off a chain reaction that leads to any of a number of symptoms, such as sneezing, nasal congestion, and coughing (respiratory or drug allergy); itchy throat, mouth, and eyes (respiratory allergy); stomachache, indigestion, vomiting, diarrhea (food allergy); itchy, reddening, swelling, or irritated skin (drug, food, and insect allergies); and swollen, stiff, and/or painful joints (food or drug allergy).

Nine of out ten cases of asthma are triggered by allergies. In people with asthma, another group of chemicals called leukotrienes cause asthmatic symptoms, which include swelling of the lung lining, spasm of the airway tubes, and excessive production of thick mucus. Leukotrienes are approximately one thousand times more potent than antihistamines in causing symptoms.

Symptoms of asthma attack include coughing, wheezing, and a feeling of tightness in the chest. Some attacks are mild, others may be life-threatening. Asthma can be triggered by a number of irritants, or even exercise. Asthma should be treated and monitored by a doctor. You may be given medication to prevent attacks, as well as a bronchodilator to open restricted airways during an attack. You should try to identify allergens so that you can avoid them.

MOST HELPFUL SUPPLEMENTS

- *Licorice* is an antiinflammatory and antiallergenic agent. Studies have found that licorice increases the half-life of cortisol, resulting in an increased antiinflammatory action of this hormone. Take any one of the following three times daily: 1 to 2 grams of powdered root, 2 to 4 milliliters of fluid extract, or 250 to 500 milligrams of solid powdered extract.

- *Nettle* relieves sneezing and itchy eyes. Take one of the following three times daily: two 300-milligram capsules or tablets or 2 to 4 milliliters of tincture.

- *Quercetin* is a bioflavonoid that is a natural antihistamine and antiasthma agent. Quercetin is often included in allergy relief combination formulas. Take 400 milligrams five to twenty minutes before each meal.

- *Vitamin C* is a natural antihistamine that has been found to prevent allergies and asthma. Studies have found that people with asthma tend to have low levels of vitamin C in their blood. In addition, double-blind controlled studies have found that as little as 1 gram of vitamin C daily can help prevent asthma attacks. Take 10 to 30 milligrams daily for every two pounds of body weight. Take in divided doses.

OTHER HELPFUL SUPPLEMENTS

- *Coenzyme Q10* helps prevent production of histamine. Take 50 to 150 milligrams daily.

- *Vitamin B6* (pyridoxine) has proven to be effective in asthma patients. In one study reported in the *American Journal of Clinical Nutrition*, participants reported a dramatic decrease in the frequency and severity of symptoms during asthma attacks when taking B6 supplements. Take 25 to 50 milligrams twice daily.

- *Selenium* levels tend to be low in asthmatics. A selenium-containing enzyme (glutathione peroxidase) is essential for reducing leukotriene formation. Take 200 to 400 micrograms daily.

- *Garlic* helps minimize allergic symptoms, including a runny nose. Garlic (and onions) inhibit the lipoxygenase enzyme, which produces an inflammatory chemical. Take one dose of odorless garlic twice a day for as long as symptoms persist.

AN OUNCE OF PREVENTION

The best way to deal with allergies is to avoid exposure to the substances that trigger outbreaks. Keeping your house clean may help prevent some reactions; special vacuum cleaners with dust-trapping filters may be beneficial. Likewise, to avoid asthma attacks, avoid the allergens or activities that trigger the attacks. If your asthma attacks stem from exercise, talk to your doctor about administering a preventive dose of asthma medication before physical activity.

CALL FOR HELP

If you suspect you have allergies, contact your doctor for advice on how to manage your condition. Allergy medication or shots may be beneficial in minimizing symptoms. Contact your doctor if you experience difficulty breathing, or if the symptoms don't clear up within one week.

Asthma should always be treated and monitored by a physician, preferably one with experience treating asthmatics. It's often hard to know when to call for help, but if you're in doubt, err on the side of caution and call your doctor or 9-1-1. As a rule, seek medical help if you must fight for breath and cannot talk, if you lean forward to get air, or if your condition does not improve within ten minutes of taking medication. Asthma is a life-threatening condition, and every attack should be treated with respect.

Alzheimer's Disease and Dementia

One of the advantages of growing older is that we accumulate the wisdom and precious memories of a lifetime. Unfortunately, Alzheimer's disease and dementia can rob us of this knowledge, which is part of what makes us who we are. More than four million Americans—including two out of three nursing-home patients—suffer from Alzheimer's disease or dementia.

Alzheimer's disease causes the gradual deterioration of the brain, resulting in a continuous decline in mental and eventually physical abilities. As the disease progresses, the nerve fibers around the hippocampus—the brain's memory center—become crossed and knotted; these neurofibrillary tangles make it impossible to store or retrieve information. In addition to this internal short circuit, the brain also experiences a drop in the concentration of neurotransmitting substances, which further breaks down the body's communication network.

Dementia (or *senile dementia*) refers to general mental deterioration, including memory loss, moodiness, irritability, personality changes, childish behavior, difficulty communicating, and inability to concentrate. Alzheimer's disease is a type of dementia.

Alzheimer's disease, dementia, and other progressive losses of mental functioning are not a normal or inevitable part of the aging process; they indicate that something has gone wrong. In a healthy person, intellectual performance can remain relatively uncompromised well into the nineties, provided the mind remains stimulated through learning. Most older people do not lose a significant amount of their mental functioning, and if they do, it is usually a result of a physical problem, such as a stroke.

Unfortunately, experts do not fully understand the cause of Alzheimer's disease and dementia, although there does appear to be a hereditary link. We do know that the brains of Alzheimer's patients tend to have high levels of aluminum, calcium, silicon, and sulfur. Some researchers speculate that the blood protein ApoE (apolipoprotein E) may play a role in the disease. The presence of this protein is determined by genetics, and several of the forms have been associated with a higher risk of Alzheimer's. It is unclear whether ApoE destroys the nerve cells in the brain or if it is involved with plaque formation.

So far there is no cure for Alzheimer's and no known way to reverse or stop progression of the disease. All therapies, both conventional and natural, aim to slow the advancement of the disease and ease symptoms.

MOST HELPFUL SUPPLEMENTS

• *Carnitine* is an amino acid that provides the brain, heart, and skeletal muscles with energy by carrying fatty acids to the cells. According to several Italian studies, carnitine supplements can slow the rate of mental decline in people with Alzheimer's disease. Red meat and dairy products are good sources of carnitine. Take 500 milligrams of L-acetyl-carnitine three times daily.

- *Coenzyme Q10* helps transport oxygen to the cells and improve brain function. Take 100 milligrams daily.

- *Garlic* has been shown in animal studies to slow down brain deterioration in laboratory rats with Alzheimer's disease. Use garlic liberally in cooking, eat up to 6 cloves of fresh garlic daily, or use odorless garlic capsules, following package directions.

- *Vitamin E* helps transport oxygen to the cells in the brain. Taken early enough, it may help to prevent Alzheimer's disease and dementia—or at least slow their progress. A study published in the *New England Journal of Medicine* followed 341 people with Alzheimer's disease who were given either a prescription drug, 2000 IU of vitamin E, or a placebo for two years. The researchers found that people taking vitamin E were 53 percent less likely to reach the advanced stages of Alzheimer's disease than those in the placebo group. (The vitamin E group fared better than the prescription drug group, too.) Take 400 IU daily; if signs of Alzheimer's disease are already present, increase the dose to 800 IU daily.

- *Vitamin B12* deficiency causes symptoms that can easily be mistaken for Alzheimer's disease or dementia. Even a mild deficiency can cause neurological disorders and memory loss. (A vitamin B12 deficiency can also be caused by pernicious anemia.) Take a multivitamin containing 10 micrograms of vitamin B12 daily.

OTHER HELPFUL SUPPLEMENTS

• *Ginkgo* alleviates anxiety, short-term memory loss, depression, inability to concentrate, and confusion when taken in the early stages. In 1997 the *Journal of the American Medical Association* published the results of a study in which mental attention and memory in Alzheimer's patients was improved by taking ginkgo. Flavonoid compounds known as heterosides are thought to be the active constituents in ginkgo most likely responsible for the increased blood flow and oxygen supply to the brain. Allow four to six weeks to see results. Take 40 to 80 milligrams three times daily.

• *DHEA* enhances memory, improves cognitive function. Take 25 to 50 milligrams for men over age fifty; 15 to 25 milligrams for women.

AN OUNCE OF PREVENTION

Scientists do not understand fully the causes of Alzheimer's disease and dementia, so there are no known steps to prevent it. Some research suggests that supplemental estrogen may help to prevent Alzheimer's disease. Preliminary research indicates that women who take estrogen for two to three years cut their risk of Alzheimer's by about 10 percent. If you have a family history of Alzheimer's disease, you may want to discuss the pros and cons of estrogen replacement therapy with your doctor.

CALL FOR HELP

If you or a loved one shows the initial signs of Alzheimer's disease—forgetfulness, short-term memory loss, disorientation, moodiness, difficulty with mathematical calculations, and difficulty finding the right words—mention your concern to your doctor. (In most patients with Alzheimer's disease, the symptoms are worse at night.) While many people fear they may be developing Alzheimer's disease, the fact that they are worried is a sign that they are not.

Anemia

Anemia is a condition in which the blood is deficient in either red blood cells or hemoglobin, which is the portion of the red blood cells that contains iron. Either of these conditions can cause a lack of oxygen to be delivered to the tissues, resulting in fatigue, weakness, and pallor.

Anemia can be caused by excessive destruction of red blood cells, excessive loss of blood, or deficient production of red blood cells. Excessive destruction may occur when red blood cells have an abnormal shape, which is seen in sickle-cell anemia and other hereditary diseases and in vitamin and mineral deficiency. Excessive blood loss may be acute (as in a trauma situation) or chronic (e.g., a bleeding ulcer, heavy menstrual flow). Deficient red-blood-cell production is the most common category of anemia, and poor nutrition is the most common cause. The three most prevalent types of nutrient-deficient anemia are those related to a deficiency of iron, folic acid, copper, or vitamin B12.

The only way to get a definite diagnosis of anemia is to get a blood test. If you suspect you have anemia, do not begin a supplement program, especially one that includes iron, until you have a diagnosis from your physician. Iron is very toxic if taken in large quantities.

Iron-deficiency anemia is seen most often in infants younger than two years old, teenage girls, pregnant women, and the elderly. Factors associated with this type of anemia include poor dietary intake of iron, reduced iron absorption or utilization, blood loss, an increased need for iron, or a combination of these situations.

Folic acid deficiency is the most prevalent vitamin deficiency on earth. The groups most likely to have folic acid anemia are alcoholics, pregnant women, and people with malabsorption conditions (e.g., Crohn's disease, celiac disease) or chronic diarrhea. In addition to anemia, a folic-acid deficiency also causes depression, diarrhea, and a swollen, red tongue.

Anemia is particularly dangerous in people who suffer from atherosclerosis (a buildup of fatty plaque in the arteries that restricts blood flow). This combination of illnesses can severely limit the delivery of oxygen to the heart, brain, legs, and other tissues, leading to severe shortness of breath, chest pain, leg pain, and even stroke.

MOST HELPFUL SUPPLEMENTS

- *Folic acid* helps prevent anemia caused by folic acid deficiency. Take 800 to 1,200 micrograms of folic acid three times daily.

- *Iron* helps prevent iron-deficiency anemia. Take 30 milligrams succinate, gluconate, or fumarate iron, twice a day between meals. If the iron causes stomach distress, switch to 30 milligrams with meals three times daily.

- *Vitamin C* helps the body absorb iron. Studies have shown that you can nearly double your absorption of iron from plant sources (such as whole grain products) by consuming vitamin C with the iron. Take 1,000 milligrams three times a day with meals.

- *Vitamin B12* helps prevent vitamin-B12-deficiency anemia. Take 2,000 micrograms sublingual three times daily for thirty days, then 1,000 micrograms methylcobalamin (the active form of B12) once per day, plus folic acid.

OTHER HELPFUL SUPPLEMENTS

- *Vitamin B5* (pantothenic acid) helps with the production of red blood cells. Take 100 milligrams daily.

- *Vitamin B6* (pyridoxine) helps with the production of red blood cells. Take 50 milligrams three times daily.

- *Nettle* is an herb rich in iron and other vitamins and minerals. For an infusion, use 1 teaspoon of powdered herb per cup of boiling water. Steep for ten to twenty minutes. Strain, and drink no more than 1 cup daily.

AN OUNCE OF PREVENTION

To prevent anemia, eat well-balanced meals, including iron-rich foods. If you have anemia, avoid foods and beverages containing caffeine, because it interferes with the body's ability to absorb iron. For the same reason, pass on iced tea, which contains tannins that also get in the way of iron absorption, as well as foods high in oxalic acids, including almonds, asparagus, beans, beets, cashews, chocolate, kale, and rhubarb.

CALL FOR HELP

If you suspect you have anemia, visit your doctor for a professional diagnosis. While most cases of anemia are caused by simple nutritional deficiencies, anemia can also be a symptom of a more serious medical problem such as cancer or leukemia.

Angina

The pain starts with constriction in the center of the chest, then radiates to the throat, back, neck, jaw, and down the left arm. You break into a sweat, struggle for breath, and feel nauseated and dizzy. You may assume you're in the throes of The Big One—a full-blown, chest-crushing heart attack—but within ten minutes or so, it's over, and the pain gradually subsides. What you've experienced is not a heart attack but an attack of *angina pectoris*.

Some three million Americans suffer from angina, a painful episode that occurs when the heart muscle does not get enough oxygen. (This is known as *myocardial ischemia*.) Most angina attacks occur when the heart, damaged by high blood pressure and coronary artery disease, is stressed by physical exertion, emotional upset, excessive excitement, or even digestion of a heavy meal. Attacks can be brought on by walking outside on a cold day, jogging to catch a bus, or hearing particularly distressing news. Angina attacks often serve as painful reminders that the heart has been damaged, and a full-blown heart attack may follow unless steps are taken to mend your ailing heart.

MOST HELPFUL SUPPLEMENTS

- *Hawthorn*, sometimes referred to as a "heart tonic," is an herb that helps in the treatment of angina and congestive heart failure by dilating the coronary arteries and improving blood circulation in the heart. Double-blind studies have shown that hawthorn extracts are effective in reducing angina attacks, as well as in lowering blood pressure. Use commercially prepared products, following package directions. You may also prepare an infusion, using 2 teaspoons of crushed leaves per cup of boiling water. Steep 20 to 30 minutes, strain, and drink up to 2 cups per day.

- *Carnitine* is a vitaminlike compound that helps the heart use oxygen more efficiently. Carnitine assists in the transportation and breakdown of fatty acids in the cells. Take up to 2 grams daily.

- *Coenzyme Q10* helps prevent the accumulation of fatty acids; it also plays a critical role in energy production in the cells. Several small studies have found coenzyme Q10 to be effective and safe in the treatment of angina. Take up to 100 milligrams daily.

OTHER HELPFUL SUPPLEMENTS

- *Magnesium* deficiency can cause spasms of the coronary arteries. In fact, people who suffer sudden, severe heart attacks often have low levels of magnesium. Magnesium helps to dilate the coronary arteries, reduce the demands on the heart, inhibit the formation of blood clots, and normalize the heart rate. Take up to 250 milligrams daily.

AN OUNCE OF PREVENTION

Since angina is caused by coronary artery disease, particularly atherosclerosis, the best way to avoid it is to keep the heart and circulatory system as healthy as possible. A healthy heart requires a healthy body. If necessary, lose weight; also exercise regularly and eat a well-balanced diet. Since smoking increases levels of carbon monoxide in the blood and limits the flow of oxygen to the tissues, smoking or being around others who do is one of the worst things you can do for your heart.

CALL FOR HELP

Angina is a symptom of serious heart disease, not a disease in its own right. If you experience chest pains associated with angina, make an appointment with a cardiologist for a complete physical.

Anxiety

Anxiety isn't all in your head: It is also in your body, often expressing itself as trembling, sweating, a pounding heart, shallow breathing, confusion, a dry mouth, and difficulty speaking. We all feel anxiety now and then, but uncontrolled anxiety or panic attacks can be debilitating and terrifying experiences, lasting from a few minutes to several hours or days. In some cases, the episodes become chronic and ongoing, interfering with a person's ability to live a happy and productive life. While anxiety attacks can strike at any age, people often experience a growing number of anxiety-producing stresses in midlife and beyond.

Along with the physical symptoms, a person experiencing an anxiety attack typically feels an overwhelming sense of terror or impending doom. Unlike fear in response to danger, anxiety is an expression of a generalized, undefined fear. The physical and emotional symptoms are so severe and terrifying that people who experience anxiety attacks often end up in the hospital emergency room, convinced they are having a heart attack.

While there is considerable debate about the causes of anxiety attacks, many experts suspect that a neurochemical imbalance can trigger both anxiety disorders and depression. (People who are prone to depression also tend to experience anxiety attacks.) While some medications and psychological treatments can be helpful in managing anxiety, natural remedies also can be helpful. Anxiety may also have potentially serious underlying causes, such as hyperthyroid disease or tumors that secrete hormones arousing the autonomic nervous system; check with your health care provider to rule out these causes if you experience repeated attacks.

MOST HELPFUL SUPPLEMENTS
- *Garlic* encourages the release of serotonin, a brain chemical that helps to regulate mood. High serotonin levels act like a tranquilizer to calm the nerves and relieve anxiety. Commercial products are available; follow package directions. For an infusion, chop 6 cloves of garlic per cup of cool water and steep for six hours; drink 1 cup a day.

- *Skullcap* helps relax the central nervous system and ease stress. To help calm the nerves, use 1 to 2 teaspoons of dried herb per cup of boiling water; steep for fifteen minutes, strain, and drink up to 3 cups daily. Commercial products are also available; follow package directions.

- *Valerian* contains chemicals known as valepotriates that have sedative properties. Commercial products are available; follow package directions. To prepare an infusion, use 2 teaspoons of powdered herb per cup of water; steep for ten minutes. Strain, and drink 1 cup before bed.

- *Calcium and magnesium* have been shown to calm the nerves and reduce anxiety when taken together. Take up to 1,500 milligrams of calcium and 750 milligrams of magnesium daily.

OTHER HELPFUL SUPPLEMENTS

- *Selenium* has been shown to reduce anxiety, depression, and fatigue. Other studies have shown that taking selenium with vitamin E and other antioxidants can improve mood and increase blood flow to the brain. Take up to 50 micrograms of selenium daily.

- *Vitamin B5* (pantothenic acid) is referred to as an "antistress" vitamin because of its ability to relieve anxiety and depression. Choose a multivitamin containing vitamin B5 or a B vitamin complex.

AN OUNCE OF PREVENTION

You may not be able to eliminate all anxiety or anxiety attacks from your life, but you may be able to minimize their frequency and intensity by getting exercise, practicing relaxation techniques, and using biofeedback to control stress. Eat a well-balanced diet, and seek professional psychological help if you need it.

CALL FOR HELP

During an anxiety attack, it can be difficult to determine whether or not you are experiencing a heart attack. If in doubt, seek medical help immediately.

Arthritis: Osteoarthritis

Over the years, the body can show signs of wear and tear. Outwardly you may sprout some gray hairs and your face may be creased with a few wrinkles. And inwardly chances are good that your joints have started to act up—the most painful sign of osteoarthritis.

Unlike other types of arthritis, which can strike suddenly, osteoarthritis usually takes its toll gradually, with the symptoms becoming worse over a number of years. Osteoarthritis afflicts, to some degree, approximately 40 million Americans—one out of every seven people—including 80 to 90 percent of people over age fifty and almost everyone over age sixty.

This degenerative joint disease involves the breakdown of cartilage and bone in the joints, especially the joints of the fingers, hips, knees, and spine. In most case, decades of use gradually damage the cartilage in the joints, causing it to harden and form bone spurs. This cartilage breakdown causes the pain, inflammation, deformity, and restriction in range of motion characteristic of osteoarthritis. Osteoarthritis can also be caused by an injury, a trauma, a physical abnormality in the joint, or a previous joint disease.

In many cases, the problem affects the synovial joint, a gel-filled capsule consisting of connective tissue that attaches one bone to another in a way that allows movement. To create a smooth hinge between the bones, cartilage covers the ends of the bones, and a membrane inside the joint secretes a lubricating gel known as synovial fluid. An outer shell or joint capsule surrounds the various components of the joint and keeps them intact. This system works remarkably well. In a healthy joint, the bones glide back and forth and allow movement without grinding or rubbing bone against bone.

In a person with arthritis, however, the joint has become damaged or diseased. There may not be enough synovial fluid in the joint, causing stiffness, or there may be too much, causing swelling around the joint. If the cartilage at the ends of the bones has worn or chipped away, bone may scrape against bone, causing additional pain.

MOST HELPFUL SUPPLEMENTS

- *Boron* plays an important role in joint health. Boron supplementation has been used in the treatment of osteoarthritis in Germany since the mid-1970s. A double-blind study published in the *Journal of Nutritional Medicine* found that among people with osteoarthritis taking 6 milligrams of boron daily, more than 70 percent improved, compared with just 10 percent of the group taking a placebo. Boron is not included in many multivitamin–mineral supplements because the federal government has not established an RDA for boron. If your daily vitamin does not include boron, take a supplement with 6 to 9 milligrams daily.

- *Copper, zinc, and vitamin A* are required for the production of collagen and cartilage. A deficiency of any one of these nutrients can cause joint degeneration. Take a multivitamin–mineral supplement that includes the RDA for copper, zinc, and vitamin A; high doses are not necessary.

- *Glucosamine and chondroitin sulfate* help with the formation of cartilage. As people age, their bodies produce less of these substances, causing cartilage to lose water and become less effective. Taking glucosamine and chondroitin supplements can reverse the effects of osteoarthritis. Glucosamine-chondroitin combination supplements are available at health food stores. The standard dose is 1,500 milligrams (divided into three doses daily). Be sure to pick up a product containing glucosamine sulfate rather than glucosamine hydrochloride, since the research on effectiveness has been performed on the sulfate form.

OTHER HELPFUL SUPPLEMENTS

- *Vitamin C* helps prevent damage to the cartilage in the joints, and it assists in the manufacture of collagen, a protein in cartilage. Without enough vitamin C, the body stops producing collagen, and the joints become compromised. Take up to 3,000 milligrams of vitamin C daily, in divided doses. Studies have also shown that vitamin E helps protect against cartilage breakdown, especially when taken in combination with vitamin C. Take 400 to 800 IU of vitamin E daily.

- *Boswellia* is used in the treatment of arthritis because of its antiinflammatory effects and because it improves the blood supply to the joints. This herb, made from the gum resin of a large branching tree in India, is available in commercial preparations; follow package directions.

- *Alfalfa* is often used to alleviate the inflammation and pain associated with arthritis. Use a commercial preparation or prepare an infusion by adding 1 to 2 teaspoons of dried herb to 1 cup of boiling water. Steep for fifteen minutes, strain, and drink up to 3 cups a day.

- *Ginger* has antiinflammatory and antioxidant properties. Studies have found ginger useful in providing pain relief, decreasing swelling and stiffness in the joints, and increasing joint mobility in people with osteoarthritis. Take 500 milligrams in capsule form daily.

AN OUNCE OF PREVENTION

To spare your joints, lose your spare tire. The more surplus weight you carry around, the greater the likelihood that you will develop osteoarthritis. Think about it: If you injured your knee, would you want to lug around a 25-pound bag of sand all day? To your knees, there's no difference between 25 pounds of fat and 25 pounds of sand. Losing the extra weight will improve your overall health in addition to reducing the stress on your weight-bearing joints. Regular exercise can also keep the joints healthy.

CALL FOR HELP

The first sign of osteoarthritis is usually joint stiffness, especially in the mornings and after long periods of rest. Other early symptoms include joint tenderness and slight swelling, cracking and creaking joints, loss of range of motion, and pain when the joint is used. Report these symptoms to your doctor during a regular physical exam.

Arthritis: Rheumatoid Arthritis

If you have rheumatoid arthritis, your immune system has turned on itself and attacked your joints and organs. Rheumatoid arthritis affects the entire body, causing chronic inflammation of many joints, as well as the skin, muscles, blood vessels, and in rare cases, organs such as the heart and lungs. If improperly treated, rheumatoid arthritis can lead to joint deformity. The disease causes the synovial membrane in the joints to divide and expand, causing inflammation and a buildup of joint fluid. Increased blood flow to the joints can cause redness and warmth.

In addition to joint problems, rheumatoid arthritis can cause fever, fatigue, weight loss, anemia, and tingling hands and feet. If the organs become involved, complications can include an enlarged spleen, irregular heartbeat, or pleurisy, an inflammation of the membrane covering the lungs. Lumps (called rheumatoid nodules) can also appear in the joints, especially in the elbow joints.

In some cases, a person has a single bout with the disease, which then disappears and never returns (monocyclic rheumatoid arthritis). Other times, a person cycles between periods of pain and periods of normal function (polycyclic rheumatoid arthritis). Most often, a diagnosis of rheumatoid arthritis means chronic pain and the need for pain management.

The diagnosis of rheumatoid arthritis can be confirmed in about 80 percent of cases by a blood test for antibodies linked to the disease. Another test, measuring the rate of sedimentation of elements in the blood, can indicate inflammation in the joints. X-rays can also confirm damage to the cartilage and bone.

The cause of rheumatoid arthritis is not fully understood. Some cases appear to be caused by a hereditary factor, but others may follow a viral infection. This serious condition plagues about seven million Americans, about three-fourths of them women. The first signs of the disease usually show up between the ages of thirty-five and forty-five.

MOST HELPFUL SUPPLEMENTS

• *Fish oils* and *omega-3 fatty acids* have been shown to relieve inflammation and the symptoms of rheumatoid arthritis. A Danish study of fifty-one people with rheumatoid arthritis showed significant improvement in stiffness and pain after 12 weeks on a daily dose of 3.6 grams of omega-3 polyunsaturated fatty acids. (The amount used in the study is equal to about one 8-ounce serving of salmon, mackerel, or herring.) The most widely available fish oil is EPA (eicosapentaenoic acid), which is available in health food stores. Another type is DHA (docosahexaenoic acid); follow package directions.

• *Selenium* and *vitamin E* can help ease the pain of rheumatoid arthritis when taken together. Many people with rheumatoid arthritis have low levels of selenium, an important antioxidant that also helps slow the body's production of inflammatory agents known as prostaglandins and leukotrienes. Selenium and vitamin E can help stave off the disease in the first place. Most good multivitamin–mineral supplements contain appropriate amounts of selenium (50 to 200 micrograms) and vitamin E (100 to 400 IU). Megadoses of these nutrients are not necessary.

- *Vitamin B5* (pantothenic acid) deficiency has been linked to rheumatoid arthritis. The lower the level of vitamin B5, the more severe the symptoms. Studies have found that raising vitamin B5 levels to the normal range helped alleviate symptoms. In one double-blind study, patients taking 2 grams of vitamin B5 experienced improvement in morning stiffness and pain, compared to a placebo group. Take up to 2 grams of vitamin B5 daily.

- *Vitamin C* helps reduce inflammation associated with rheumatoid arthritis. Without enough vitamin C, the body stops producing collagen, and the joints become compromised. Take up to 3,000 milligrams of vitamin C daily, in divided doses.

OTHER HELPFUL SUPPLEMENTS
- *Zinc* deficiencies are common in people with rheumatoid arthritis. Take up to 45 milligrams of zinc daily. Some multivitamin–mineral supplements contain this level of zinc.

AN OUNCE OF PREVENTION
The cause of rheumatoid arthritis is not fully understood; at this point there are no specific recommendations on how to avoid the disease.

CALL FOR HELP
Rheumatoid arthritis usually shows up first as pain when moving a joint, especially early in the morning. It usually occurs first in the wrists and knuckles, or the knee and ball of the foot, though it can affect any joint of the body. If you experience chronic or severe joint

pain, consult with your doctor to have your condition diagnosed.

Atherosclerosis and Arteriosclerosis

The phrase "hardening of the arteries" refers to two very similar conditions: arteriosclerosis, which is the accumulation of calcium deposits on the inside of artery walls; and atherosclerosis, which is the accumulation of fatty substances (plaque) on the artery walls instead of calcium. In both cases, blood circulation is restricted as it moves through hardened, narrow arteries. The restricted blood flow causes blood pressure to rise, cells to die, and often, a heart attack or stroke.

Atherosclerosis and arteriosclerosis typically develop in three steps. First, the arteries develop tiny tears due to the powerful contractions of the heart, especially in someone who has high blood pressure. Next, cholesterol in the blood sticks to the tears, slowly hardening into plaque, causing the arteries to become less flexible. Finally, these deposits narrow the arterial passages, reducing the blood supply to the heart muscle and other parts of the body.

The heart muscle is so efficient at extracting oxygen from the blood that many people develop severe coronary disease before any symptoms appear. In fact, the vessels can be 70 to 90 percent blocked before any symptoms occur—and often a heart attack is the first warning sign that something is wrong.

When it involves the coronary arteries, atherosclerosis causes heart attack. When it blocks blood flow to the brain, atherosclerosis causes stroke. And when it affects the arteries of the legs, it causes peripheral vascular disease.

Hardening of the arteries is usually associated with poor diet and lifestyle habits. Some of the main factors that contribute to arteriosclerosis and atherosclerosis are a diet that is high in fat and cholesterol and low in fiber, smoking, lack of adequate exercise, stress, being overweight, diabetes, and high blood pressure. During the early stages of arteriosclerosis and atherosclerosis, which can begin in childhood, most people notice no symptoms. Some experience a dull, cramplike pain in their legs, ankles, hip, or buttocks, which may indicate partially blocked blood vessels.

MOST HELPFUL SUPPLEMENTS

- *Folic acid* helps lower levels of homocysteine. The body uses amino acids found in cow's milk and red meat to form homocysteine, an amino acid that helps create arterial lesions. Folic acid helps reduce homocysteine levels and lowers the risk of heart disease. Take 800 micrograms daily, along with vitamin B12.

- *Garlic* contains several sulfur compounds that block the biosynthesis of cholesterol. Garlic also helps expand the blood vessel walls, increasing blood flow and lowering blood pressure. Another chemical in garlic, ajoene, helps prevent blood clots. Use garlic liberally in cooking; eat up to six cloves of fresh garlic daily; or use commercially prepared odorless garlic capsules, available in health food stores.

- *Hawthorn*, known as the "heart herb," enhances cardiac output, in addition to opening up the peripheral vessels to improve overall circulation. Commercial products are available; follow package directions.

For an infusion, use 2 teaspoons of crushed leaves per cup of boiling water. Steep twenty minutes; strain, and drink up to 2 cups daily.

- *Inositol* helps to lower cholesterol. Take 1,000 milligrams daily under a doctor's supervision.

- *Vitamin E* helps prevent free radical damage, in addition to helping the body maintain cardiac and smooth muscle. It also keeps one particular form of LDL cholesterol from oxidizing and forming plaque deposits on the arteries. Take 400 to 800 IU daily.

- *Vitamin C* is essential for cholesterol metabolism. It is responsible for the excretion of excess cholesterol from the body, and it helps the body balance "good" and "bad" lipids, one of the biggest factors in heart disease. Take up to 3,000 milligrams daily, in divided doses.

OTHER HELPFUL SUPPLEMENTS

- *Calcium* is essential for blood clotting; it also plays a role in maintaining blood pressure. Take up to 1,500 milligrams daily.

- *Vitamin A* helps maintain elasticity in the tissues. Studies have shown that people who eat large amounts of beta-carotene (a form of vitamin A) have a significantly decreased mortality rate from cardiovascular disease, compared with those who don't eat much beta-carotene. The vitamin A in a multivitamin–mineral supplement should be sufficient.

- *Quercetin* is a bioflavonoid that lowers the risk of heart disease. Take 35 milligrams daily.

- *Selenium* appears to protect against heart disease and stroke; the rates for heart disease are highest where selenium intake is lowest. Take 200 micrograms daily.

AN OUNCE OF PREVENTION

The major risk factors for heart disease and circulatory problems include smoking, high blood pressure, obesity, high blood cholesterol levels, lack of exercise, and a family history of the disease.

CALL FOR HELP

Some people learn they have heart disease when they experience their first heart attack or attack of angina. Since atherosclerosis and arteriosclerosis do not have clear warning signs in their early stages, your best way to detect the disease is to have complete annual physicals.

Athlete's Foot

Athlete's foot doesn't discriminate. It afflicts both couch potatoes and toned jocks with equal vigor. Athlete's foot, *tinea pedis*, is a fungal infection that causes inflammation, itching, blisters, scaling, and burning of the feet and toes. The fungus thrives in warm, moist environments, including the inside of shoes and gym locker rooms. It is the same fungus that causes jock itch and ringworm.

Athlete's foot is contagious and can be passed along on towels, shower floors, and so on. But most people get athlete's foot from wearing sweaty socks and shoes. People who have taken antibiotics for two weeks or longer are susceptible to athlete's foot because the drugs kill the good bacteria that help prevent foot fungal infection.

MOST HELPFUL SUPPLEMENTS

- *Garlic* helps destroy fungus and fight infection. Take two 500-milligram tablets three times daily and place tiny slivers of fresh garlic in your shoes for several days.

- *Tea tree oil* (used topically) is an excellent antifungal herbal treatment. If the fungus is widespread, soak your feet for ten minutes in a soothing foot bath consisting of one quart of warm water and 10 to 15 drops of tea tree oil. Dry the feet thoroughly, especially between the toes, then apply undiluted tea tree oil directly to the affected areas and leave on the skin.

- *Goldenseal* (used topically) contains two alkaloids that have antifungal properties. Soak your feet up to three times daily in a medicinal tea made with goldenseal. Use ½ ounce of powdered goldenseal and one pint of boiling water. Simmer for fifteen minutes; strain and add to a foot tub or bowl of tepid water. Bathe your feet and then dry thoroughly.

OTHER HELPFUL SUPPLEMENTS

- *Acidophilus* helps restore good bacteria that help fight fungal infections. Take one of the following forms three times daily: 1 tablespoon liquid extract; 1 or 2 capsules, tablets, or softgels; or 2 tablespoons powder in cool liquid.

- *Zinc* boosts the immune system and fights fungus. Take 50 milligrams daily.

AN OUNCE OF PREVENTION

If you are susceptible to athlete's foot, take strides to keep your feet as clean and dry as possible. Wear white cotton socks that allow feet to breathe, and wash the socks in chlorine bleach after each wearing to kill any fungus that might be present. If you tend to have sweaty feet, change your socks several times daily, and whenever possible, wear sandals or open-toe shoes to allow air to reach your feet. And, of course, wear waterproof slippers or sandals in locker rooms and public showers to avoid contact with the fungus.

CALL FOR HELP

Athlete's foot can usually be defeated by consistent and aggressive home treatment. The fungal infection can be accompanied by a bacterial infection, however, which could require medical attention. If the condition persists or gets worse after one month of treatment, call your doctor.

Bronchitis and Pneumonia

Just when you think you're getting better, the coughing begins. Your cold and flu symptoms finally subside, but what follows is a hacking cough and sudden fever—clear signs of bronchitis.

Bronchitis involves inflammation, irritation, or infection of the breathing tubes, or bronchi, in the upper part of the lungs. It can be triggered by a bacterial or viral infection or by an allergic reaction to something you have inhaled, such as pollens, cat dander, or dust. As the lining of the bronchi swell, mucus builds up, causing an irresistible urge to cough and clear the throat.

Bronchitis comes in two forms. Acute bronchitis has a viral cause in 90 percent of cases and is characterized by a hacking cough, fever, chills, tightness in the chest, and yellow, green, or white phlegm that usually appears twenty-four to forty-eight hours after the cough begins. Chronic bronchitis is more serious and usually develops in people who are overweight, sedentary, and exposed to smoke. Symptoms include a persistent cough that results in yellow, green, or white phlegm that lasts for at least three months of the year and for more than two consecutive years. People with chronic bronchitis are at high risk of developing heart disease and more serious lung disease.

Most bouts with bronchitis last only a week or so. If symptoms persist, call your doctor, who can determine if the condition is caused by a bacterial infection, in which case antibiotics may be needed.

Pneumonia is an infection or irritation of the lungs. Both conditions usually follow a bout of common cold, flu, or other respiratory illness. Pneumonia can be caused by bacteria, fungi, or viruses, and usually develops in people who have a weakened immune system, such as hospitalized patients or elderly individuals with chronic medical conditions. Symptoms of pneumonia include fever, chills, cough, muscle aches, swollen lymph glands, fatigue, sore throat, chest pains, and difficulty breathing.

Pneumonia is actually an infection of the alveoli, the tiny air sacs in the lungs, which become inflamed and then fill with mucus and pus. When infected, the alveoli cannot do their job of transferring oxygen from the lungs to the blood, so the person feels tired and out of breath. If the body is denied enough oxygen, the lips, fingernails, and toenails may turn somewhat blue.

Pneumonia can also be caused by stroke, alcoholism, smoking, kidney failure, sickle-cell disease, malnutrition, foreign bodies in the respiratory passages, chemical irritation, and allergies. Pneumonia is a serious illness that should be treated by a physician, though you can also use natural remedies to help manage your symptoms.

MOST HELPFUL SUPPLEMENTS

- *Beta-carotene* protects the lungs and soothes the mucous membranes. Deficiency of beta-carotene and vitamin A increases the susceptibility to respiratory infections and pneumonia. Take 10,000 IU daily.

- *Vitamin C* helps reduce inflammation. Take 3,000 milligrams daily in divided doses.

- *Zinc* helps bronchial healing. Proper immune function and tissue repair depends on adequate levels of zinc in the body. Suck on zinc lozenges, following package directions for dosage information.

OTHER HELPFUL SUPPLEMENTS

- *Echinacea* enhances the immune system. Take any one of the following forms three times daily: 0.5 to 1 gram of dried root (as a decoction); 2 to 4 milliliters tincture; 2 to 4 milliliters fluid extract; 150 to 300 milligrams dry powdered extract.

- *Garlic* is a natural antibiotic. Take 2 odorless tablets with meals.

- *Vitamin E* heals tissues and improves breathing. Take 400 IU twice daily.

AN OUNCE OF PREVENTION

You can't prevent every cold, but you may be able to prevent some colds from turning into bronchitis by treating them as soon as the first symptoms develop. If you have allergies, take steps to minimize respiratory problems by avoiding exposure to allergens.

If you haven't had a vaccine against pneumonia and you're in good health and over age sixty-five, call your doctor to schedule an appointment to be immunized. Also talk to your doctor about the vaccine if you're under age sixty-five but have an increased risk of developing pneumonia because you have Hodgkin's disease, an immune disorder, or heart, kidney, liver, or lung disease. A single injection of the vaccine provides lifelong immunity against about 90 percent of the types of bacteria that cause pneumococcal infections in the United States.

CALL FOR HELP

Most cases of bronchitis can be treated at home and run their course in ten days or so. However, call your doctor if you experience a fever of 103 degrees or higher, extreme lethargy, wheezing, or difficulty breathing.

In the early stages, it's difficult to tell the difference between a bad cold and pneumonia. But if a cough following a cold lingers for more than a week, if you cough up green, yellow, or brown sputum, or if you experience rapid or labored breathing, call your doctor.

Bursitis and Tendonitis

Inflammation of the bursae—fluid-filled sacs that cushion the bones, ligaments, and tendons when they move against each other—is known as bursitis. Inflammation in or around the tendon is known as tendonitis. Both of these conditions involve pain, inflammation, and swelling and can restrict movement, but they do differ somewhat. Bursitis commonly affects the shoulders, elbows, hips, knees, and the joints in the feet and hands. The symptoms are especially apparent when lifting, stretching, or whenever a joint is moved repetitively or beyond its normal range of motion.

In tendonitis, the tendons become inflamed when they are overused, either repetitively or occasionally. The areas most often affected are the shoulder, wrists, heel, and elbow (tennis elbow). Other symptoms include stiffness and, in some cases, swelling, tingling, and numbness. Although acute tendonitis usually heals within a few days to 2 weeks, it may become chronic, in which case calcium salts typically deposit along the tendon fibers.

MOST HELPFUL SUPPLEMENTS

- *Boswellia* helps reduce inflammation. Take 150 milligrams three times daily.

- *Bromelain* has been shown to reduce inflammation, regardless of cause. Several studies have confirmed its efficacy in treating bursitis and tendonitis. Take 250 to 750 milligrams three times daily between meals.

- *Vitamin C* plays an important role in the repair of injuries. Vitamin C assists with collagen production, which is essential for the formation of tendon and bursal tissues. Take 1,000 milligrams daily.

OTHER HELPFUL SUPPLEMENTS

• *Sulfur* relieves inflammation and pain. Take 500 to 5,000 milligrams daily. Start with a low dosage and increase by 500 milligrams every few days until you notice results.

• *Zinc* helps produce collagen, which builds tissue. Take 20 milligrams for tendonitis.

AN OUNCE OF PREVENTION

Proper stretching and warming up before exercise can help prevent some cases of bursitis and tendonitis.

CALL FOR HELP

After an injury, follow the **RICE** approach to first aid: **R**est the injured part, **I**ce the area to decrease swelling, **C**ompress the area to limit swelling, and **E**levate the part above the heart level to increase fluid drainage out of the injured area. The ice and compression should be applied for thirty minutes, followed by fifteen minutes without ice and compression to allow recirculation. For serious injury, contact your physician immediately. Also call your doctor if you experience severe pain, loss of function, or pain that persists for more than two weeks.

Cancer

Consider the odds: Every day our bodies produce more than 500 billion new cells. Every once in a while an error occurs, and our bodies form defective cells. This can be the beginning of cancer.

Cancer develops when oncogenes (the genes that control cell growth) are transformed by a carcinogen, or cancer-causing agent. In most cases, the immune system identifies and destroys these aberrant cells before they multiply. But when the system breaks down, these fast-growing cancer cells reproduce, forming a tumor and invading healthy tissue. These tumors zap the body of nutrients and interfere with the tasks performed by the healthy tissue.

While cancer can develop at any age, it tends to affect older people more than younger ones because they have been exposed to more carcinogens over a longer period of time. In many cases, it takes years or decades for cancer-causing agents to do damage; other cancers grow and spread rapidly.

While all cancers involve the uncontrolled growth of cells, the word cancer refers to more than 100 different diseases. There are four main categories of cancer: carcinomas of the skin, mucous membranes, glands, and other organs; leukemia of the blood; sarcomas of the muscles, connective tissues, and bones; and lymphomas of the lymphatic system.

Not all tumors are cancerous: Benign (noncancerous) tumors do not spread and infiltrate surrounding tissues; malignant (cancerous) tumors do spread or metastasize through the blood vessels and lymph system to other areas of the body, where new tumors grow. Areas of the body where malignant tumors most commonly develop are the bone marrow, breasts, colon, liver, lungs, lymphatic system, ovaries, pancreas, prostate gland, skin, stomach, and uterus.

Cancer is the second most common cause of death in the United States, after heart disease. Despite its prevalence, the exact cause of cancer remains a mystery, although some experts argue that environmental factors—such as exposure to tobacco smoke, radiation, asbestos fibers, and toxic wastes—cause about 80 percent of all cancers. In addition, some people may inherit a greater sensitivity to carcinogens and a greater propensity to develop cancer.

MOST HELPFUL SUPPLEMENTS
- *Antioxidants*—vitamins A (beta carotene), C, E, and selenium—help protect against many forms of cancer. Low blood levels of beta-carotene have been linked to an increased risk of different types of cancer, especially breast, bladder, colon, and lung cancer. Similarly, vitamin C has been shown to protect against breast cancer in postmenopausal women, as well as colon and stomach cancer. A deficiency in vitamin E has been linked to breast cancer. Vitamin E may also help to prevent stomach cancer and other cancers of the gastrointestinal tract by inhibiting the conversion of nitrates in foods to nitrosamines, which are potential carcinogens. Broad-based epidemiological studies have shown that selenium may protect against cancers of the breast, colon, and lung. Selenium inactivates peroxides in the cells, which can result in tissue damage. Take 10,000 IU beta carotene, 5,000 to 10,000 milligrams vitamin C, 1,000 IU vitamin E, and 200 micrograms selenium daily.

- *Coenzyme Q10* inhibits the formation of free radicals, unstable oxygen molecules that can damage the cells and cause cancer. Take 100 milligrams daily.

OTHER HELPFUL SUPPLEMENTS
- *Garlic* has been found to help prevent stomach cancer and to enhance the immune system. Use it liberally in cooking. Take 2 capsules three times daily.

- *Green tea* boosts the immune system. Drink up to 5 cups tea daily, preferably with meals, or take 500-milligram capsules daily.

- *Calcium* has been shown to reduce the risk of colon cancer as well as to reduce the formation of polyps (precancerous lesion). It is thought that calcium binds to bile acids in the gut and neutralizes their cancer-causing effects. One study found that a daily intake of just 375 milligrams of calcium (approximately the amount in one eight-ounce glass of milk) was associated with a 50 percent reduction in the rate of colon cancer, while consuming 1,200 milligrams of calcium was linked to a 75 percent decrease in colon cancer. Take up to 1,500 milligrams daily.

- *Arginine* is a nonessential amino acid that stimulates the release of growth hormone produced by the pituitary gland in the brain. Animal studies have found that arginine helps inhibit the growth of tumors. Studies with human blood cells show that arginine increases the production of immune cells that interfere with tumor growth. Commercial products are available; follow package directions.

- *Folic acid* plays a key role in building DNA, the complex compound that forms our genetic blueprint. Observational studies show that people who get higher than average amounts of folic acid from their diets or supplements have lower risks of colon and breast cancer. (This could be especially important for those who drink alcohol, since alcohol blocks the absorption of folic acid and inactivates circulating folate.) An interesting observation from the

Nurses' Health Study is that high intake of folic acid blunts the increased risk of breast cancer seen among women who have more than one alcoholic drink a day. Take 400 micrograms of folic acid daily.

• *Lycopene* is a photochemical found in fruits and vegetables in the carotenoid family. Lycopene gives fruits and vegetables their reddish color; it is found in red peppers, ruby red grapefruit, and tomatoes. Studies have shown a link between low blood levels of lycopene and increased risk of bladder and pancreatic cancer. In one study, men who consumed the greatest amounts of lycopene (6.5 milligrams per day) in their diets showed a 21 percent lower risk of prostate cancer compared with those eating the least. Commercial products are available; follow package directions.

AN OUNCE OF PREVENTION

While not all cancers can be avoided, taking steps to minimize your lifetime exposure to carcinogens can help lower your risk of developing some cancers. Don't smoke; don't drink heavily; eat a low-fat, high-fiber diet; maintain your optimal weight; avoid unnecessary x-rays; check your home for radon gas; and limit your exposure to sun and wear sunscreen. In addition, visit your doctor regularly for routine screening tests.

CALL FOR HELP

While many cancers grow silently in the body for months or years before making themselves known, contact your doctor immediately if you detect any cancer warning signs, such as a lump under the skin, a persistent cough, bloody sputum, difficulty swallowing, chronic indigestion, discharge from the nipple, bleeding between periods, change in bowel habits, blood in stool or urine, headaches and visual disturbances, excessive bruising, nosebleeds, unexplained weight loss, abdominal pain, a sore that does not heal, or change in a wart or mole.

Canker Sores

Canker sores (*aphthous stomatitis*) are small, painful ulcers that develop on the inside of the mouth, alone or in small clusters. They form when mouth acids and digestive enzymes eat away the soft tissue of the mouth. Canker sores can be caused or aggravated by certain foods—such as nuts, coconut, and chocolate—as well as by stress or minor mouth injuries. Unlike cold sores, canker sores cannot be spread from person to person.

As many as 50 percent of Americans get canker sores each year. For some reason, certain unfortunate people are far more prone than others to experience outbreaks, which usually last 4 or 5 days. Women are twice as likely to develop them, especially before their menstrual period begins. Most canker sores go away after a few days, but they recur in many people.

MOST HELPFUL SUPPLEMENTS
- *Licorice* has antiviral and antibacterial properties, and it can ease the irritation of a canker sore. Take 200 milligrams powdered deglycyrrhizinated licorice dissolved in 200 milliliters warm water. Swish in the mouth for two minutes three to four

times daily. Or chew a 380-milligram chewable tablet twenty minutes before meals.

- *Myrrh* promotes healing. Take 200 to 300 milligrams extract or 4 milliliters tincture with 2 to 3 ounces warm water. Swish in the mouth for several minutes and swallow. Repeat two to three times daily.

- *Vitamin B complex* promotes healing and maintains B vitamin balance. Take 50 milligrams three times daily.

OTHER HELPFUL SUPPLEMENTS

- *Acidophilus* reduces soreness and helps restore bacterial balance. Chew four tablets three times daily until the sores heal.

- *Dandelion* helps canker sores heal, and it prevents their recurrence. Take 500- to 1,000-milligram capsules three times daily for six weeks.

- *Echinacea* promotes healing. Take 4 milliliters liquid extract echinacea swished in the mouth for two to three minutes three times daily. Swallow the extract for added protection.

- *Vitamin C* promotes tissue healing. Take 3,000 milligrams daily in divided doses.

AN OUNCE OF PREVENTION

Since no one knows exactly what causes canker sores, it's impossible to offer surefire ways of preventing them. However, you can reduce the number of sores by avoiding sharp, hard-bristled toothbrushes and other objects that can injure or irritate the lining of the mouth. In addition, a daily helping of yogurt can help keep the lining of the mouth healthy.

CALL FOR HELP

Canker sores should heal within two weeks. If a sore lingers or recurs, contact your doctor or dentist. You may need an antibiotic or a doctor may need to cauterize the sore. In addition, a sharp end to a tooth or another dental problem could aggravate a canker sore and not give it a chance to heal.

Carpal Tunnel Syndrome

In the computer age, carpal tunnel syndrome is an increasingly common disorder characterized by pain, weakness, numbness, tingling, and burning in the wrist and fingers that often radiates to the forearm and shoulder. It is caused by compression of the median nerve, which controls the movements of the fingers and thumbs. The median nerve runs from the forearm to the fingertips and lies between the ligaments and bones of the wrist.

People who perform a lot of repetitive movement of the hand, such as typists, assembly-line workers, keyboard operators, cashiers, and carpenters, are among those most likely to suffer with the condition. It can also follow a more serious injury to the wrist, or it can be caused by another condition that causes swelling of the tissues in the wrist, such as rheumatoid arthritis, diabetes, or hypothyroidism.

MOST HELPFUL SUPPLEMENTS

- *Bromelain* relieves pain and inflammation. Take 250 to 750 milligrams twice daily between meals.

- *Vitamin B6* (pyridoxine) helps relieve pain associated with carpal tunnel syndrome in some—but not all—patients. Generally, vitamin B6 helps some patients, especially when it is combined with supplements of vitamin B2 and other B vitamins. In any case, treatment with vitamin B6 is safe and recommended before opting for surgery. Take 25 milligrams three to four times daily.

- *Vitamin B complex* enhances effectiveness of vitamin B6. Take 50 milligrams once daily.

AN OUNCE OF PREVENTION

Many cases of carpal tunnel syndrome can be prevented by wearing a specially made splint for the hand and wrist, using a wrist pad when typing or keyboarding, and taking frequent breaks to exercise the wrist and hand. Studies have shown the stretching exercises can help relieve carpal tunnel syndrome. Hold your arms out with your palms down, then flex your wrists so that your fingers point to the ceiling, then down to the floor. Repeat for up to five minutes.

CALL FOR HELP

Consult your doctor if you continue to experience pain after one month of self-treatment. In some cases, surgery is required to ease the pain.

Cataracts

If you live long enough, the chances are good you will develop cataracts. This degenerative eye problem affects more than 90 percent of people over age sixty-five. In fact, it is the leading cause of blindness in the United States.

Cataracts occur when the crystalline lens of the eye grows cloudy or opaque. The lens, located behind the pupil and the iris, connects to the muscles surrounding the eye and flexes and bends as the eye focuses. As the eye ages, the lens gradually increases in size, weight, and density. Cataracts form when the eye loses the ability to maintain appropriate concentrations of sodium, potassium, and calcium within the lens, affecting vision. If your lens were a window, a cataract could seem like either a spot or a film of steam clouding the glass and obscuring your vision.

In most cases, the mineral imbalance in the lens stems from damage caused by free radicals, due to exposure to ultraviolet light or low-level radiation from x-rays. Cataracts can also be caused by disease, particularly diabetes, injury to the eye, a congenital defect, the use of certain steroids, or exposure to German measles during fetal development.

Cataracts don't always behave in a predictable manner. They can develop quickly or slowly; they can affect one eye or both; they can progress at different rates from one eye to the other. Cataracts aren't obvious in the early stages without the use of special instruments (the whitish film on the surface of the eye doesn't show up until the cataracts are quite severe), so regular eye exams are essential.

MOST HELPFUL SUPPLEMENTS

- *Antioxidants*—vitamin A (beta-carotene), C, E, and selenium—help prevent cataracts. Take 10,000 IU beta-carotene, 3,000 milligrams vitamin C in divided doses, 400 IU vitamin E, and 400 micrograms of selenium daily.

- *Bilberry* contains bioflavonoids, which aid in the removal of chemicals from the retina of the eye. During World War II, Royal Air Force pilots snacked on bilberry jam sandwiches before flying night missions to sharpen their night vision. In one human study, bilberry extract plus vitamin E stopped progression of cataract formation in forty-eight out of fifty patients with age-related cataracts. Take 40 to 80 milligrams extract three times daily.

OTHER HELPFUL SUPPLEMENTS

- *Copper* promotes healing and retards growth of cataracts. Take 3 milligrams daily.

- *Vitamin B2* (riboflavin) deficiency has been linked to cataract formation. (About one out of three people over age sixty-five are riboflavin deficient.) Take riboflavin supplements of up to 10 milligrams daily as part of a B-complex formula.

- *Zinc* promotes healing and retards growth of cataracts. Take 50 milligrams daily.

AN OUNCE OF PREVENTION

To minimize your risk of developing cataracts, protect your eyes from ultraviolet light by avoiding direct sunlight and by wearing dark sunglasses and a wide-brim hat when outdoors. Since free radicals appear to be the leading cause of cataracts associated with aging, eating a diet rich in antioxidants can help protect the eyes.

Of course, regular eye exams are necessary to detect and treat cataracts in their early stages. Have your eyes tested by an ophthalmologist at least every five years.

CALL FOR HELP

Cataracts tend to impair vision gradually, making it difficult to notice slight changes in vision. Regular eye exams are necessary to catch the disease in its early stages. Symptoms of more advanced cataracts include dull, fuzzy vision; glare in bright light (the cataract scatters the light before it reaches your retina); double vision; and changes in color vision (the cataract emphasizes yellows and reduces violets and blues). In addition, if you find yourself suddenly able to read without your regular reading glasses, get your eyes checked. This is a sign that your lenses are changing shape.

Chronic Fatigue Syndrome

Chronic fatigue syndrome (CFS) is a term that describes a group of symptoms, including persistent and recurrent fatigue, low-grade fever, swollen lymph nodes, muscle weakness, headache, muscle and joint pain, sore throat, depression, and loss of concentration. Diagnosis can be difficult, since the symptoms of chronic fatigue syndrome are similar to those of other conditions, especially fibromyalgia and multiple chemical sensitivity disorder.

Chronic fatigue syndrome was formally defined by the Centers for Disease Control in 1988 when diagnostic criteria were established. Despite this more recent recognition of the condition, chronic fatigue syndrome apparently was identified first in the 1860s and has since then been called various names, including chronic-mononucleosis-like syndrome, yuppie flu, and postinfection neuromyasthenia.

The cause of chronic fatigue syndrome is uncertain. One possibility is the Epstein-Barr virus (EBV), which is a member of a larger group of herpes viruses. EBV is suspect because, similar to other viruses in the herpes group, it remains dormant in the body until the immune system becomes weakened. At that time the virus can become active and cause symptoms typical of CFS. Other organisms also have been named as possible causes of CFS, including herpesvirus-6, brucella, and enterovirus.

There are many ways the immune system can become impaired: stress, poor diet, insufficient sleep, smoking, alcohol or drug use. A deficiency of nearly any nutrient can make the immune system susceptible to invasion by harmful organisms that can greatly weaken the body and cause fatigue.

Chronic fatigue alone can have many causes and not necessarily be associated with CFS. Some of the major causes of chronic fatigue include the presence of yeast infection, diabetes, heart disease, rheumatoid arthritis, lung disease, chronic pain, cancer, liver disease, and multiple sclerosis. Depression, high levels of stress, food allergies, anemia, as well as use of antihypertensive drugs, tranquilizers, birth control pills, antiinflammatory medications, and antihistamines are also linked with chronic fatigue.

MOST HELPFUL SUPPLEMENTS

- *Ginseng (Siberian)* strengthens the immune system, which is often out of balance in people with chronic fatigue. In one double-blind study, 36 healthy subjects received either 10 milliliters of Siberian ginseng extract or a placebo for four weeks. The group receiving the ginseng demonstrated significant improvement in several markers of immune system strength. Take any of the following three times daily: dried root, 2 to 4 grams; 10 to 20 milliliters tincture; 2 to 4 milliliters fluid extract; 100 to 200 milligrams dried powdered extract.

- *Magnesium* relieves fatigue and muscle pain. An underlying magnesium deficiency can cause chronic fatigue. Low magnesium levels have been found in many people with chronic fatigue. Double-blind studies of people with chronic fatigue have shown that magnesium supplementation significantly improves energy levels, emotional state, and pain. Take magnesium citrate or aspirate, 200 to 300 milligrams three times daily.

- *Licorice* is an antiviral that can help relieve chronic fatigue symptoms, especially among people with adrenal insufficiency or abnormally low blood pressure. Take any one of the following three times daily; 2 to 4 grams powdered root; 2 to 4 milliliters fluid extract; 250 to 500 milligrams dried powdered extract.

OTHER HELPFUL SUPPLEMENTS

- *Coenzyme Q10* boosts the immune system. Take 75 milligrams daily.

- *Astragalus* strengthens the immune system. Take 250 to 500 milligrams dried root capsules two to three times daily with meals or water, or 3 to 5 milliliters ($\frac{1}{8}$ to $\frac{1}{2}$ teaspoons) tincture or extract three times daily.

- *Acidophilus* restores "good" bacteria in those with candida infection, a common cause of chronic fatigue syndrome. Take 1 to 2 billion CFU daily.

AN OUNCE OF PREVENTION

Since the precise cause of chronic fatigue syndrome has not been identified, no specific steps can be taken to avoid it. In general, maintaining a strong immune system helps ward off the condition.

CALL FOR HELP

Chronic fatigue syndrome can be difficult to diagnose and treat. If symptoms persist after one month of self-treatment, seek medical help with a doctor who specializes in the treatment of patients with chronic fatigue.

Colds and Flu

There's no getting around it: you're going to get sick—often—as you face a lifetime of building a strong immune system. More than two hundred different viruses cause colds and flu, and you have to build immunity to them one by one. That's why kids catch six to ten colds a year, and adults come down with two to four annually.

Technically speaking, a cold is a viral infection of the upper respiratory tract, which includes the nose, throat, sinuses, and bronchial tubes. When confronted with the viral invaders, the nose and throat release chemicals to stoke up the immune system. The affected cells produce prostaglandins, which stimulate inflammation and lure the infection-fighting white blood cells. The body temperature rises to boost the immune response and more nasal mucus is produced to trap and wash away viral particles.

Both the common cold and the flu involve general malaise, upper-respiratory-tract congestion, sneezing, fever, sore throat, and runny nose. But when you have the flu, you know you're up against more than the common cold. The flu involves the cold symptoms, plus chills, fatigue, red eyes, and sore muscles. Nausea and vomiting may also occur.

While your doctor can prescribe medication to shorten the duration of the flu if infected, in most cases all you need to do is treat the symptoms and allow the virus to run its course. Antibiotics won't kill viruses, though they may be called in if you develop pneumonia, ear infection, or another secondary infection. Natural remedies for cold and flu are geared to boost the immune system and help the body shed the virus, which ultimately results in symptom relief. Along with the treatment options mentioned below, rest and consumption of large amounts of fluids (herbal tea, water, broth, and diluted juices) are recommended to help eliminate the virus.

Is It a Cold or the Flu?

Influenza and the common cold share certain characteristics, but there are ways to distinguish the two

- Fever: Characteristic of the flu; rare with a cold.
- Headache: Common with the flu; rare with a cold.
- Aches: Severe with the flu; slight with a cold.
- Fatigue: Severe and lingering with flu; minor with a cold.
- Runny nose: Minor with the flu; characteristic of a cold.
- Sore throat: Possible with flu; common with a cold.
- Cough: Common with both flu and cold, but more severe with the flu.

MOST HELPFUL SUPPLEMENTS

- *Astragalus* is a traditional Chinese medicine used to treat the common cold and other viral infections. Studies have found astragalus to be effective in reducing the duration and severity of cold symptoms. Research indicates that astragalus stimulates the white blood cells to destroy invading organisms and to enhance the production of interferon, a natural compound that fights viruses. Take 250 to 500 milligrams dried root capsules two to three times daily with food or water, or 3 to 5 milliliters (⅛ to ½ teaspoon) extract or tincture three times daily.

- *Echinacea* enhances the immune system and helps fight viruses. More than 350 studies have examined the immune-enhancing properties of echinacea; while not all of the studies have had positive results, the majority have suggested that echinacea helps reduce the duration and severity of cold symptoms. Take any one of the following forms three times daily: 0.5 to 1 gram dried root (as a tea); 2 to 4 milliliters tincture; 2 to 4 milliliters fluid extract; 150 to 300 milligrams dried powdered extract.

- *Vitamin A* heals inflamed membranes and boosts the immune system. Take 15,000 to 25,000 IU for up to four days.

- *Vitamin C* helps fight viruses and reduce inflammation. Since 1970, more than twenty double-blind studies have shown that vitamin C helps decrease the duration and severity of cold symptoms. Take 1,000 milligrams every two hours (as tolerated) daily.

- *Zinc* boosts the immune system and helps fight certain common cold viruses. Dissolve 1 lozenge (15 to 25 milligrams each) of elemental zinc under the tongue every three hours for the first three days, then one every four hours for up to another four days. Do not take zinc for longer than seven days. For flu, dissolve 1 lozenge every two hours for up to one week.

OTHER HELPFUL SUPPLEMENTS
- *Garlic* enhances the immune system and helps fight infection. Take 2 capsules 3 times daily.

- *Licorice* tea soothes a sore throat and helps relieve coughs. Use 1 teaspoon of dried herb per cup of boiling water. Steep for ten minutes, strain, and drink up to three cups daily.

AN OUNCE OF PREVENTION

You can't avoid every cold, but you may be able to decrease your odds of catching every one that's being passed around if you practice good hand-washing habits. During cold and flu season, take steps to maintain a healthy immune system by eating right and getting plenty of rest.

Flu shots (vaccinations against the current flu strain) help prevent flu, but the shot itself can cause flulike symptoms. Talk to your doctor about whether you should have a flu shot.

CALL FOR HELP

Most colds don't require a doctor's visit, but you should call your physician if you develop shortness of breath, a fever that lasts more than four days, or sinus pain and congestion that lingers more than 10 days.

Sometimes the flu leads to serious complications, such as pneumonia or encephalitis, so contact the doctor if you develop a cough that lasts more than 1 week or a stiff neck.

Constipation

It could be caused by something you ate—or something you didn't eat. It could be a side effect of a new medication, or your body's rebellion to stress or travel. Whatever the cause, constipation can be difficult—and painful—to live with.

Constipation is not defined by frequency of bowel movements. In fact, when it comes to bowel movements, there is no single definition of "normal." Some people have three bowel movements a day, others have three per week, but neither group is considered to have a problem as long as the stools are soft enough to pass without difficulty. In general, the body excretes waste in eighteen hours (with a high-fiber diet) to forty-eight hours (with a low-fiber diet).

Constipation involves straining to pass dry, hard stools; it can also include gas pains, bloating, headache, and indigestion. In severe cases, it can cause the digestive system to grind to a halt, resulting in an impacted bowel, as well as hemorrhoids and varicose veins.

Constipation is usually the result of an insufficient intake of fiber and fluids, although occasionally it is caused by certain drugs or supplements such as iron tablets, antidepressants, and some painkillers. Constipation is also common during pregnancy.

MOST HELPFUL SUPPLEMENTS
- *Aloe vera* is a potent stimulant laxative that helps form soft stools and promotes healing. It helps heal and cleanse the digestive tract, while helping the formation of soft stools. Drink ½ cup aloe vera juice in the morning and evening, or take a 50- to 200-milligram latex tablet or softgel once a day up to ten days as needed.

- *Psyllium* is a bulk-forming laxative. Sometimes referred to as "Mother Nature's laxative," psyllium seeds are the active ingredient in many over-the-counter laxative preparations. Take 1 to 2 rounded teaspoons after each meal with a full glass of water. Commercial products are available; follow package directions.

- *Folic acid* deficiency has been found to be present in people with chronic constipation; the condition often resolves when supplements are used. Take 400 micrograms of folic acid daily.

• *Magnesium*: Magnesium is frequently used as a laxative. Magnesium sulfate, magnesium hydroxide (milk of magnesia), and magnesium citrate are so-called saline-type laxatives that draw fluid from the tissues and serum into the intestine. The result is stimulation of the intestine to contract. (Overuse of these laxatives over long periods of time can cause magnesium toxicity.) Use a commercial product; follow package directions.

OTHER HELPFUL SUPPLEMENTS

• *Vitamin E* helps heal an irritated colon. Take 400 IU daily.

• *Acidophilus* supplements boost levels of beneficial bacteria in the intestine, which can help with digestion and relieve constipation. Take 1 teaspoon twice daily, or follow package directions.

AN OUNCE OF PREVENTION

Constipation can be prevented through changes in diet and exercise habits in most cases. Drink plenty of fluids (at least eight 8-ounce glasses of water daily) and eat plenty of high-fiber foods (such as oat bran, fruits, and vegetables with skin). Regular vigorous exercise can also stimulate the bowels and relieve constipation.

CALL FOR HELP

Constipation should be treated promptly. Not only are the symptoms unpleasant, but chronic constipation can lead to hemorrhoids, indigestion, and diverticulitis. If constipation lingers after more than four days of self-treatment, contact your physician.

Chronic or recurring constipation may be an indication of a more serious condition, such as irritable bowel syndrome, colorectal cancer, multiple sclerosis, diabetes, or Parkinson's disease. If you suffer with chronic constipation, see your physician.

Dandruff

Dandruff is characterized by flaking and scaling of dead skin from the scalp. This condition may be itchy and occasionally is accompanied by a scaly rash. The telltale white flakes of dandruff affect everyone to some degree, because every body is constantly shedding skin cells and growing new ones. The problem is remarkably common: as many as one in five people has a scalp that sheds enough cells to have them show up as flakes or scales.

When the proteins and fats in the scalp function properly, they maintain circulation of water and oils and keep the skin and hair healthy. Dandruff develops when the sebaceous glands in the scalp cannot adequately assimilate proteins and fats. Occasionally dandruff is caused by an imbalance in the kidneys or liver, two organs that are responsible for eliminating toxins from the body. Dandruff doesn't hurt the hair, it doesn't cause baldness, and it doesn't mean the hair is greasy or unclean.

MOST HELPFUL SUPPLEMENTS
• *Tea tree oil* prevents flaking and infection. Massage a few drops into the scalp daily, or look for a shampoo that contains tea tree oil.

- *Vitamin B complex* helps break down fatty acids. Take 100 milligrams twice daily with meals.

OTHER HELPFUL SUPPLEMENTS

- *Evening primrose oil* relieves inflammation. Take two 500-milligram capsules three times daily.

- *Zinc* promotes protein metabolism (scalp is mainly protein). Take 5 tablets dissolved in mouth daily for one week.

AN OUNCE OF PREVENTION

Wash your hair often with nondrying shampoos, such as products designed for children or those with conditioners. Steer clear of soap-based shampoos, as well as those containing fragrances. If you use hair-styling products, avoid those containing alcohol to prevent further drying of the scalp.

CALL FOR HELP

Contact your doctor if you experience hair loss, or if your scalp becomes painful, inflamed, or really itchy. Other skin problems, such as ringworm and other fungal infections, can look a bit like dandruff, even though they require different treatment.

Depression

At some point in their lives, most people experience depression. Though many people confuse the two, depression isn't the same thing as sadness. While we all feel sadness in response to situations– the death of a loved one, the loss of a job, a divorce, or some other disappointment—depression is characterized by ongoing feelings of worthlessness, pessimism, sadness, and lack of interest in life. With clinical depression, these feelings linger for weeks or months and ultimately become incapacitating.

Depression can be either a short-term, minor problem or a lifelong, life-threatening illness. Some people inherit a tendency to develop depression due to their brain chemistry. Other times the illness is brought on by physical conditions, such as stroke, hepatitis, chronic fatigue syndrome, chronic stress, thyroid disease, menopause, alcoholism, or drug abuse—or even by the lack of natural light during the darker winter months. Some drugs, including over-the-counter antihistamines as well as many others, can cause depression, too.

Whatever the cause, most cases of depression involve an imbalance of neurotransmitters, or chemical messengers in the brain. While depression was once considered a shameful psychiatric condition, most experts now recognize that it usually has both physical and psychological triggers. It is an organic illness involving physical biochemical changes in the body, so without help the person cannot "snap out of it." While counseling and professional care can be crucial in recovery, a number of natural remedies may also prove useful.

MOST HELPFUL SUPPLEMENTS

- *St. John's wort* has long been used as a mood enhancer and antidepressant. More than thirty double-blind studies involving more than 2,000 patients with mild to moderate depression have found St. John's wort to be effective in the treatment of depression. It contains a chemical (hypericin) that blocks the action of monoamine oxidase (MAO) in the body. (MAO inhibitors are a common class of prescription antidepressant drugs.) Most people begin to report effects within the first two weeks, but it can take three to four weeks for the herb to have a significant effect on some people. St. John's wort has been found to be as effective or more effective than conventional antidepressant drugs in treating mild to moderate depression, but it is not as effective in treating severe depression. For mild to moderate depression, take 250 to 300 milligrams two to three times daily of standardized form 0.3% hypericin. NOTE: While taking this herb, do not take amphetamines, narcotics, diet pills, asthma inhalants, decongestants, or cold and hay fever medication.

- *SAMe* has been found to be one of the most effective natural antidepressants. Studies have shown SAMe to be more effective than some prescription antidepressants, with fewer side effects. Commercial products are available; follow dosage recommendations on the package.

- *Folic acid* has a mild antidepressant effect, probably due to its function as a methyl donor, which helps increase levels of serotonin in the brain. Folic acid supplementation in depressed patients taking antidepressant drugs was found to enhance the antidepressant action of the drugs. Take 100 to 200 milligrams three times daily.

OTHER HELPFUL SUPPLEMENTS
- *Niacin* improves circulation in the brain. Take 100 milligrams three times daily.

- *Oats* and oatmeal have mild antidepressant and mood-elevating properties. Eat oatmeal cereal, or take a commercially prepared oat extract.

- *Vitamin C* is necessary for the synthesis of neurotransmitters in the brain. Take up to 3,000 milligrams daily, in divided doses.

- *Vitamin B12* deficiency creates mood disturbances and depression. Correcting an underlying vitamin B12 deficiency can result in a dramatic improvement in mood. Take 1,000 micrograms daily.

AN OUNCE OF PREVENTION

Sadness is part of life, but clinical depression doesn't have to be. Debilitating depression should be treated by a mental health professional. But you may be able to prevent some types of depression—as well as generally elevate your mood—by getting regular exercise and rest and by sharing your feelings with someone you trust. Be aware that depression and sadness can also be a side effect of many medications, including over-the-counter antihistamines. If you suspect your mood changes are drug induced, talk to your doctor.

CALL FOR HELP

It can be difficult to tell the difference between clinical depression and common sadness. But there are certain warning signs.

- changes in sleep (either insomnia or sleepiness)

- changes in weight and eating habits (either weight gain or weight loss)

- loss of sexual desire or libido
- chronic fatigue or tiredness
- low self-esteem or self-worth
- loss of productivity at work, home, or school
- inability to concentrate or think clearly
- withdrawal or isolation
- loss of interest in activities that were once enjoyable
- anger or irritability
- trouble accepting praise or affirmation
- feeling slow, every activity takes supreme effort
- apprehension about the future
- frequent weeping or sobbing
- thoughts of suicide or death

These are all warning signs and diagnostic criteria for depression. If you or a loved one experience three or more of them for two weeks or longer, contact a doctor or mental health professional for help. Don't try to treat serious depression by yourself. And if you or someone you're concerned about feels suicidal, immediately seek help from a specialist or a twenty-four-hour hotline; look in the phone book under "Suicide Prevention."

Diabetes

Diabetics must live well-balanced lives: Every day they must carefully watch their blood-sugar levels. If their levels rise too high and stay there too long, they risk damage to the nerves and blood vessels, which can cause a number of health problems, including blindness, infections, kidney problems, stroke, and heart disease. But if their blood-sugar levels drop too low—even for a few minutes—they can become confused and even lose consciousness.

Normally the pancreas regulates this delicate balance of sugar in the bloodstream. But the 14 million Americans with *diabetes mellitus* cannot properly convert food (especially sugar) into energy, either because their bodies do not produce enough insulin (a hormone produced in the pancreas to regulate blood-sugar levels) or because their bodies don't properly use the insulin they do produce. Instead, diabetics must monitor their blood-sugar levels, adjusting their diet and exercise—or their oral medication and insulin injections—to meet these changing conditions.

There are two basic types of diabetes: the more severe form, known as Type I, insulin-dependent, or juvenile diabetes (about 15 percent of cases); and Type II, non-insulin-dependent, or adult-onset diabetes (about 85 percent of cases).

- Type I diabetes usually strikes sometime between the onset of puberty and age thirty. It is caused by damage to the insulin-producing cells in the pancreas. For some reason, it affects males more often than females.

- Type II diabetes usually occurs in middle-aged and older people, especially those who are overweight, although in recent years it has been showing up in children and young adults with poor diet and exercise habits. Losing as little as ten or fifteen pounds helps control Type II diabetes in most cases. With Type II diabetes, the pancreas produces insulin, but the sugar remains in the bloodstream. This more subtle version of the disease often goes undetected, until complications arise. Ultimately, up to 60 percent of Type II diabetics need supplemental insulin.

Both Type I and Type II diabetes seem to have a genetic component as well. Other possible causes include an immune response, following a viral infection that destroys the cells in the pancreas. Diabetes can also follow other diseases, such as thyroid disorders, inflammation of the pancreas, or problems with the pituitary gland. In addition, about 5 percent of women develop diabetes when pregnant, though the symptoms usually disappear after the baby is born.

NOTE: Diabetes is a condition for which there are a multitude of natural remedies that have been proven effective, so the number of options listed here is greater than for most of the other conditions. It is important to work with a medical professional when taking any of these supplements, as their use may significantly decrease or eliminate your need for antidiabetic medication.

MOST HELPFUL SUPPLEMENTS

• *Chromium* improves glucose tolerance. Chromium makes insulin about ten times more efficient at processing sugar, so less insulin is needed to do the job. Unfortunately, levels of chromium in the body tend to decrease with age. Nearly twenty controlled studies have demonstrated a positive effect for chromium in the treatment of diabetes. Most of the studies were performed in patients with Type II, non-insulin dependent diabetes. Take 200 micrograms of chromium daily.

- *Fenugreek* helps to regulate insulin. Studies have shown that this herb can reduce urine sugar levels by 50 percent. Researchers have used 15 to 100 grams fenugreek powder daily to treat people with non-insulin-dependent diabetes. Do not take fenugreek without the guidance of a medical professional.

- *Biotin* enhances insulin sensitivity and improves the utilization of blood sugar. This response is thought to be the result of an increase in the activity of the enzyme glucokinase, which is involved in the utilization of blood sugar by the liver. In one study, 8 milligrams of biotin twice daily resulted in significant lowering of fasting blood sugar levels and improved blood glucose control on Type I diabetics. In a study of Type II diabetics, similar effects were noted with 9 milligrams of biotin daily. Take 9 to 16 milligrams daily, under a doctor's supervision.

- *Magnesium* improves insulin production. Many diabetics have a deficiency in magnesium; supplements (even at low doses) tend to minimize complications related to the disease. Take 300 to 400 milligrams daily.

- *Manganese* levels in diabetics tend to be about half the levels of normal individuals. Take 30 milligrams daily.

OTHER HELPFUL SUPPLEMENTS
- *Inositol* helps maintain normal nerve function. Diabetic neuropathy is the most common complication

of long-term diabetes. Much of the decrease in nerve function is due to loss of inositol from the nerve cells. Take 500 milligrams twice daily.

- *Coenzyme Q10* assists in carbohydrate metabolism. Take 120 milligrams daily.

- *Vitamin B6* (pyridoxine) improves glucose tolerance. People with diabetic neuropathy have been shown to be deficient in vitamin B6 and to benefit from supplementation. Take 1,800 milligrams of vitamin B6 daily in the form of pyridoxine alpha-ketoglutarate.

- *Vitamin C* is transported into the cells using insulin. Many diabetics develop vitamin C deficiencies in the cells, even if they consume enough in their diets. Failure to address the problem with additional supplementation can lead to health problems, including poor wound healing, high cholesterol levels, and a depressed immune system. Take 1,000 to 3,000 milligrams daily.

- *Vitamin E* improves glucose tolerance and helps the body maintain normal blood-sugar levels. Take 400 IU daily.

- *Garlic* helps lower blood-sugar levels. Eat three to six cloves of garlic daily, or use a commercially prepared product.

- *Ginseng* has long been used to treat diabetes. Double-blind studies have shown that ginseng can improve glucose control and increase energy among people with Type II diabetes.

AN OUNCE OF PREVENTION

While not all cases of diabetes can be prevented, many can. Maintain a healthy weight (most diabetics weigh thirty to sixty pounds more than they should), eat a low-fat, high-fiber diet, and of course, exercise regularly. Studies have shown that vigorous exercise can lower the risk of developing Type II diabetes by one-third. In fact, many experts consider exercise the most effective way to prevent non-insulin-dependent diabetes.

CALL FOR HELP

Diabetes can be difficult to detect. In fact, only about half of all diabetics know they have the disease. The symptoms of Type I diabetes—excessive thirst, frequent urination, dry mouth, blurred vision, and frequent infections—often develop rapidly. The signs of Type II diabetes—thirst, drowsiness, obesity, fatigue, tingling or numbness in the feet, blurred vision, and itching—often go unrecognized for years before being properly diagnosed. If you experience any of the warning signs, contact your doctor immediately. A diabetes screening should be part of your annual physical exam.

Diarrhea

The digestive system isn't very discriminating. When confronted with a toxin, the system shifts into overdrive and simply clears itself of anything—and everything—in the way. The result: Diarrhea, which involves frequent, loose, or watery stools, and abdominal cramping.

Diarrhea is not a disease, but a symptom of another

underlying problem. It can be caused by viruses, bacteria, protozoa, food poisoning, food allergy, lactose intolerance, or dietary imbalances, such as too much fiber. The condition is not only unpleasant, but it can be dangerous if the fluids lost by the body are not replenished in a timely fashion. Most bouts of diarrhea clear up in twenty-four to forty-eight hours, but the potentially life-threatening complications of dehydration can show up in a matter of hours in a child, so it demands immediate attention and treatment.

A deficiency of folic acid or zinc has been linked with bouts of diarrhea, as has excessive (more than 2,000 or 3,000 milligrams) amounts of vitamin C. Consumption of foods that contain the natural sweetener called sorbitol is also known to cause diarrhea. Nondietary causes include use of certain drugs, such as antibiotics (especially tetracycline), antacids that contain magnesium salts, or laxatives that contain magnesium, phosphate, or sulfate. People with medical conditions such as Crohn's disease, ulcerative colitis, hepatitis, or cancer typically can experience diarrhea as well.

Because diarrhea causes the body to lose a great deal of water and essential nutrients, individuals should replace them by drinking lots of herbal tea, fruit and vegetable juices, vegetable broth, or electrolyte-replacement drinks, which supply chloride, potassium, and sodium. Milk and dairy products should be avoided.

MOST HELPFUL SUPPLEMENTS
- *Acidophilus* replaces "good" bacteria in the digestive tract. Take 1 teaspoon powder in distilled water twice daily.

- *Brewer's yeast* relieves infectious diarrhea. Take capsules or tablets three times daily for two weeks.

- *Folic acid* reduces recovery time and heals the intestinal walls. Take 5,000 micrograms three times daily for two or three days.

OTHER HELPFUL SUPPLEMENTS

- *Garlic* kills parasites and bacteria that cause some cases of diarrhea. Take 2 capsules three times daily.

- *Goldenseal* helps cure cases of diarrhea caused by parasites. Take 4 to 6 grams powdered root capsules daily or 4 to 6 milliliters liquid extract daily.

AN OUNCE OF PREVENTION

Meticulous hand-washing habits may prevent some outbreaks of diarrhea caused by food poisoning or bacterial infection. If you suspect a food allergy or lactose intolerance due to recurring diarrhea after eating certain foods, monitor your diet carefully to identify the offending foods.

CALL FOR HELP

With diarrhea, you need to look out for signs of dehydration. Symptoms include infrequent urination, dark or concentrated urine, loss of skin elasticity, and lethargy. Call the doctor if you experience dehydration, bloody diarrhea, severe abdominal cramping, or if the diarrhea persists for more than two or three days.

Diverticulitis

Diverticulitis is a disease of the colon (large intestine) in which grape-sized sacs (known as diverticula) develop in the walls of the intestine and extend out into the surrounding body cavities. When food particles become lodged in the sacs, the colon becomes inflamed and infected. Symptoms of diverticulitis include fever, severe pain in the lower abdomen, nausea, diarrhea, and constipation. Some patients have no symptoms at all for years before having an attack with pain. In severe cases, the disease progresses to a potentially fatal stage called peritonitis.

Individuals at high risk of developing diverticulitis include those who are obese or who have a poor diet (low fiber, high fat), family history of the disease, gallbladder disease, or coronary artery disease. It is estimated that 50 percent or more of Americans over age fifty have diverticulitis to some degree. Because the walls of the intestines weaken as a person ages, the condition is more common in older people. Diverticulitis is common in people who suffer from frequent constipation.

Once diverticula develop they do not go away. The pouches themselves do not cause problems, but they are prone to infection. People can have diverticulosis— they have the pouches—without having diverticulitis— inflammation of the pouches.

MOST HELPFUL SUPPLEMENTS

- *Acidophilus* restores "good" bacteria in the digestive tract. Take 1 teaspoon three times daily on an empty stomach.

- *Aloe vera* helps to gently clean the intestinal tract. (It also helps prevent constipation.) Take 1 teaspoon gel after meals. Don't take more, because it can act as a laxative.

- *Psyllium* supplies fiber, which softens the stools and cleans the intestines. Take 1 teaspoon in water once or twice daily.

OTHER HELPFUL SUPPLEMENTS

- *Garlic* promotes healing. Take 2 capsules with meals.

- *Vitamin B complex* promotes healing. Take 100 milligrams three times daily.

AN OUNCE OF PREVENTION

Eating a diet high in fiber and getting regular exercise can promote good bowel habits and minimize your risk of developing diverticulitis. Take steps to avoid constipation (see page 211), a condition common in people who suffer from diverticulitis.

CALL FOR HELP

If you develop fever and severe abdominal pain, contact your doctor immediately. You may have an infection within the digestive tract, which could require prompt medical attention.

Ear Infections

Ear infections are painful conditions of the inner or outer ear, which tend to recur and typically affect children more than adults. In fact, approximately 95 percent of children experience at least one ear infection by age six.

The ear has three main parts: The outer ear (where the sound waves are caught and directed into the ear canal toward the eardrum), the middle ear (where the sound waves vibrate and bounce around three tiny bones), and the inner ear (where a tiny spinal structure turns the vibrations into nerve impulses to the brain).

Many ear problems begin with the Eustachian tube, a small tube that connects the middle ear and the throat to equalize air pressure and avoid rupturing the eardrum. Unfortunately, the Eustachian tube can also allow bacteria to travel from the throat and nose into the middle ear, causing an infection known as *otitis media*. People with shorter Eustachian tubes tend to develop more ear infections than others because it's easier for the bacteria to invade the middle ear. When the Eustachian tubes become infected and inflamed, fluid builds up in the middle ear, decreasing hearing and causing a stuffy feeling in the head. As children grow, the angles of their Eustachian tubes change, allowing fluids to drain more readily. That's why most ear infections occur in children between the ages of 6 months and 3 years.

Alas, infections aren't the exclusive domain of the middle ear. Swimmer's ear (or *otitis externa*) is an infection of the outer ear that usually occurs in the summer when people are in the water a lot (though it can afflict nonswimmers as well). The infection shows up as itching or tingling outside the ear, sometimes with a yellowish discharge. If your ear hurts when you gently pull it and wiggle it, chances are good you have an outer-ear infection. Symptoms include fever, temporary loss of hearing in the affected ear, and a discharge from the ear.

Researchers have found a definite link between food allergies and recurrent ear infection, especially chronic *otitis media*. More than half of children with recurrent ear infections have food allergies. When the culprit foods (milk, wheat, and eggs are the common allergens) are eliminated, more than 75 percent of children improve significantly. Other factors associated with ear infection include exposure to secondhand smoke, smoke from wood-burning stoves, and being bottle-fed or having been breast-fed for less than four months.

MOST HELPFUL SUPPLEMENTS

- *Echinacea* kills bacteria and boosts the immune system. Take 1 dose three times a day for three days, followed by 1 dose once a day for four more days.

- *Vitamin C* boosts immune response. Take 3,000 milligrams daily in divided doses.

- *Zinc* boosts immune response. Take 10 milligrams as a lozenge three times daily for five days.

OTHER HELPFUL SUPPLEMENTS

- *Manganese* eliminates the deficiency associated with ear infection. Take 10 milligrams daily.

- *Garlic* helps kill the bacteria associated with ear infections. Take one dose of odorless garlic daily for 5 days, or follow package directions.

- *Acidophilus* kills infectious bacteria and helps maintain beneficial bacteria in the intestine, which will be destroyed if you take antibiotics to treat an ear infection. Take 1 dose daily for one week after taking an antibiotic, or follow package directions.

AN OUNCE OF PREVENTION

Some, but not all, ear infections can be prevented. To minimize your risk of middle-ear infection, blow your nose gently (blowing too hard can force bacteria into the ear). Outer-ear infections are caused by moisture in the outer ear, so keeping your ears bone-dry can help prevent infection. After bathing, hold a hair dryer eighteen inches from your ear and blow warm (not hot) air into the ear. It should take only a minute or two to dry any remaining moisture.

CALL FOR HELP

If you experience loss of hearing or suffer from a fever, contact your doctor. You may need antibiotics to treat the infection. If you experience ear pain, then notice a yellowish or bloody discharge from the ear, call your physician. Your eardrum may have ruptured. The eardrum will heal (in fact, the rupture is part of the body's healing process), but your doctor may need to prescribe antibiotics to speed the healing and prevent additional infection.

Eczema

Eczema is the skin's way of telling you there's a problem. Those red, itchy patches announce to the world that you have dry skin caused by an allergy or local irritation. When you scratch the area, the skin gets thick, scaly, and rough; sometimes the skin oozes and becomes crusty. The condition is unpredictable and can strike any part of the body, but most outbreaks occur on the creases of the elbows and knees, behind the ears, and on the face and wrists.

Eczema (*atopic* or *contact dermatitis*) can appear at any age; it affects approximately 2 to 7 percent of Americans. Eczema is at least partially caused by allergies, proven by the fact that all people with eczema test positive on allergy tests and most improve when they consume a diet that eliminates common food allergens such as eggs, wheat, milk, and peanuts. It tends to be triggered by common allergies, such as hay fever, asthma, sensitivity to perfumes, chemical reactions, and food allergies. Other characteristics shared by most people with eczema are dry, thickened skin that has very limited capacity to hold water; an overgrowth of bacteria, especially *Staphylococcus aureus*, which is found in 90 percent of patients; and a tendency for the skin to thicken when rubbed or scratched.

Nutritional and supplemental treatment of eczema focuses on preventing the release of excess histamine that is associated with allergic reactions and providing nutrients that offer antiinflammatory and antiallergenic benefits. Deficiencies often seen in people with eczema include zinc and gamma-linolenic acid (GLA), a fatty acid. People with eczema do not have the ability to properly process fatty acids.

MOST HELPFUL SUPPLEMENTS

- *Evening primrose oil* restores GLA levels, which tend to be low in people with eczema. Take two 500-milligram capsules three times daily. Evening primrose oil can also be applied topically once a day to reduce inflammation and redness.

- *Licorice* is both antiinflammatory and antiallergenic. Take one of the following three times daily: 1 to 2 grams powdered root, 2 to 4 milliliters fluid extract, or 250 to 500 milligrams dry powdered extract.

- *Chamomile* has been found to be very effective in reducing the inflammation and itching of eczema, probably because of its naturally antiinflammatory chemical flavonoids. Chamomile is available commercially; follow package directions.

- *Zinc* restores low zinc levels usually seen in patients with eczema; it also promotes tissue repair. Take 45 to 60 milligrams daily, reduce to 30 milligrams when eczema clears.

OTHER HELPFUL SUPPLEMENTS
- *Green tea* is both an antihistamine and antiallergenic. Take 200 to 300 milligrams three times daily.

- *Quercetin* is a bioflavonoid that is an antihistamine and antiallergenic agent. Take 400 milligrams twenty minutes before meals.

- *Oatmeal* baths can help soothe itchy skin. Use a commercial preparation; follow package directions.

AN OUNCE OF PREVENTION
If you have sensitive skin, you need to avoid everything that irritates the skin or causes allergic reactions. Wear cotton instead of wool, silk, or synthetics. Keep your skin moisturized by taking quick showers (or baths) and frequent applications of moisturizer. If you swim in a chlorinated pool, rinse thoroughly to remove all traces of harsh chemicals.

CALL FOR HELP
The first time you develop eczema, your condition should be diagnosed and treated by a doctor. Subsequent outbreaks can be treated at home, except when you develop an open sore that shows sign of infection. Look for yellowish discharge or pus, swelling, and red streaks in the area of the site.

Fibrocystic Breast Disease

About half of all women have fibrocystic breast disease, a term used to describe several benign (noncancerous) conditions that affect the breast. Women with fibrocystic breast disease often experience tenderness or pain as well as lumps in their breasts caused by the enlargement of glandular tissue. The lumps, or cysts, move freely in the breast, are either firm or soft, and may change in size. Pain results as the cysts fill with fluid and the tissue around them becomes thick, placing pressure on the surrounding area. This fluid is normally reabsorbed by the breast tissue, but as a woman ages the ability of the lymph system to absorb the fluid decreases, and cysts remain. The cysts are most painful before menstruation.

Fibrocystic breast disease appears to be caused by an imbalance in the ratio of estrogen to progesterone, as well as cyclic changes in the levels of other hormones. The condition can be exacerbated by compounds known as methylxanthines, which are found in coffee, tea, cola, chocolate, and medications that contain caffeine.

Fibrocystic breast disease is often a complaint associated with premenstrual syndrome (PMS). If you have symptoms of PMS as well as fibrocystic breasts, the treatment options in the entry on PMS on page 302 may be more helpful. If fibrocystic breast disease is your primary complaint, try the options below.

MOST HELPFUL SUPPLEMENTS
- *Evening primrose oil* may reduce the size of the cysts in the breasts. Take 2 capsules three times daily.

- *Vitamin B6* (pyridoxine) helps to regulate fluid retention. Take 50 milligrams three times daily.

- *Vitamin E* helps minimize symptoms. Several double-blind studies have confirmed the usefulness of vitamin E in treating fibrocystic breast disease. Take 1,000 IU daily of emulsion for one month.

OTHER HELPFUL SUPPLEMENTS

- *Acidophilus* promotes the excretion of excess estrogen. Take 1 tablespoon liquid extract, 1 or 2 capsules, tables, or softgels; or 2 tablespoons powder in cool liquid.

- *Chaste berry* helps balance hormones. Take 40 drops liquid extract or one capsule daily in the morning.

- *Germanium* eases pain and promotes tissue oxygenation. Take 100 milligrams daily.

AN OUNCE OF PREVENTION

Avoid consuming coffee, tea, cola, chocolate, and medications that contain caffeine. These trigger foods can provoke symptoms of fibrocystic breast disease.

CALL FOR HELP

Some women become alarmed by the presence of lumps in their breasts due to a well-placed fear of breast cancer. It is important to perform monthly breast self-exams and to visit your doctor regularly for an annual physical and breast exams. Women over age forty should have regular mammograms, at their doctor's recommendation. If you detect a suspicious lump, contact your doctor to rule out the possibility of a tumor.

Fibromyalgia

An estimated three to six million people, most of whom are women between the ages of twenty-five and forty-five, have fibromyalgia—a condition with uncertain causes and no known cure. The primary symptom is severe pain in the muscles, joints, and ligaments, and it is usually accompanied by disturbed sleep patterns, insomnia, and fatigue. Other symptoms include chronic headache, swollen lymph nodes, irritable bowel syndrome, swollen joints, numbness or a tingling sensation, and depression.

People with fibromyalgia have chronically low levels of the hormone serotonin. This deficiency causes the sensation of pain to be much exaggerated as well as the sleep problems that affect most people who have this disease. A magnesium deficiency is also associated with fibromyalgia. Stress exacerbates the pain, and low-impact exercise is recommended to help ease symptoms.

MOST HELPFUL SUPPLEMENTS

- *5-HTP* is converted by the body into serotonin. Take 50 to 100 milligrams three times daily.

- *Magnesium* helps to burn fat and produce energy, which is crucial to combat fibromyalgia. A study conducted at the USDA's Grand Forks Human Nutrition Research Center found that postmenopausal women with low magnesium levels had less energy and did not burn fat efficiently as women with normal magnesium levels, making physical exertion more difficult. Another study found that people taking daily magnesium supplements experienced significant improvement in the number and severity of tender points. Take 150 to 250 milligrams three times daily.

• *St. John's wort* works with 5-HTP and magnesium to boost serotonin levels. Take 300 milligrams three times daily.

OTHER HELPFUL SUPPLEMENTS

• *Cayenne* (capsaicin) relieves pain; apply lotion containing 0.025 to 0.075 percent capsaicin to affected areas.

• *SAMe* has been found to help ease joint pain and other symptoms of fibromyalgia. At least four clinical studies have found SAMe to produce excellent results in reducing both the number of trigger points and to improve overall mood. Take up to two 500-milligram capsules daily.

AN OUNCE OF PREVENTION

At this time little is known about the causes of fibromyalgia, so there are no recommendations on how to prevent it.

CALL FOR HELP

If you experience chronic fatigue and pain, consult your doctor and schedule a complete physical. Your doctor should be asked to rule out other serious illnesses (such as leukemia or cancer) that could cause similar symptoms.

Flatulence

Flatulence (from the Latin word *flatus*, meaning blowing) is a normal part of the digestive process. In deed, the average person releases nearly 1 quart of gas each day. Gas forms when undigested food passes from the small intestine into the colon, where bacteria feast on the remains and give off noxious gases in the process. It can also be caused by swallowed air and food allergies. Once it is trapped, gas has only two means of escaping—through belching or flatulating.

Foods most likely to cause gas include beans, dairy products, fried foods, and glutenous grains. Beans and grains can be made more easily digestible if you soak them overnight and then cook them in fresh water. Foods to avoid if you are prone to flatulence include fried foods, sugar, hydrogenated fat (especially in snack junk foods), and very cold liquids. Combining spice or acidic foods with dairy product or sugary foods also can cause flatulence.

MOST HELPFUL SUPPLEMENTS

- *Alfalfa* contains chlorophyll, which aids in digestion. Drink up to three cups of tea daily or take 1 tablet 3 times daily.

- *Acidophilus* helps with digestion and can ease gas pain by encouraging the growth of the "good" bacteria necessary for a healthy digestive system. Take 1 dose twice a day, following dosage information on the product label.

- *Ginger* aids in digestion and minimizes the impact of gas-producing foods, such as beans. Add 1 ounce of grated ginger root to 16 ounces of boiling water and let steep fifteen minutes. Drink 2 to 3 cups daily. Alternatively, take ¼ to ½ teaspoon of liquid extract daily.

OTHER HELPFUL SUPPLEMENTS

- *Peppermint tea* enhances digestion and acts as an antiflatulent. Drink one cup daily.

- *Goldenseal* aids in digestion. Take 25 drops tincture three times daily.

- *Chamomile* minimizes the impact of gas-producing food. Prepare tea using 1 teaspoon of herb per cup of boiling water. Strain and drink 2 to 3 cups daily. Alternately, consume ¼ to ½ teaspoon of liquid extract.

AN OUNCE OF PREVENTION

To some degree, you can control the frequency of flatulent outbursts by regulating your diet and eating habits. Eat slowly to minimize the amount of air you swallow. If flatulence is a frequent problem, look for signs of lactose intolerance.

CALL FOR HELP

A bit of gas and occasional gas pains are nothing to worry about. However, a stubborn stomachache or abdominal pain (especially on the lower right-hand side) should be taken seriously, since it could be a sign of appendicitis. If in addition to gas pains you have a fever, nausea, vomiting, or diarrhea, contact your health care provider to rule out a more serious illness.

Gallstones

Gallstones are unpredictable: These rock-hard structures can be smaller than a pea or larger than an egg. They can sit quietly and not cause any trouble or they can block the bile duct and become inflamed, resulting in severe pain in the upper right abdomen, often accompanied by fever, nausea, and vomiting.

Gallbladder disease is serious business. Untreated, gallbladder inflammation (also called *cholecystitis*) can be life threatening. If the bile duct is blocked by a gallstone, bile can back up in the system, causing the skin and whites of the eyes to turn yellow with jaundice and the urine to turn dark brown.

These complications arise from problems with the concentration of bile, the yellowish substance the body uses to digest fat. The liver produces bile (which consists of cholesterol, bile salts, and lecithin, among other substances), and any surplus is stored in the gallbladder, a small organ nestled under the liver. If the bile in the gallbladder becomes too concentrated, the cholesterol can crystallize, forming gallstones. While 80 percent of all gallstones are composed primarily of cholesterol, they can also be formed of pure bile or mixtures of bile, cholesterol, and calcium.

Women tend to develop gallbladder disease more often than men; as many as one out of four women over age fifty-five has gallstones. Extra pounds put you at extra risk of developing gallstones; people more than 20 percent overweight double their risk of developing gallbladder disease. Other risk factors include eating a high-fat and high-sugar diet, rapid weight loss, lack of exercise, diabetes, hypertension, and estrogen replacement therapy (the estrogen increases the cholesterol levels in the bile). Native Americans are also at high risk.

Chronic gallbladder disease sometimes necessitates the surgical removal of the gallbladder (a procedure known as a cholecystectomy), but many people can control the disease by making dietary changes and using nutritional supplements.

MOST HELPFUL SUPPLEMENTS

* *Milk thistle* protects against gallstone formation. Take 600 milligrams (standardized to 70 to 80 percent silymarin content).

* *Peppermint* helps dissolve gallstones. Take 1 to 2 capsules (0.2 milliliters oil per capsule) three times daily between meals.

* *Psyllium* helps lower cholesterol levels, reducing the risk of gallstone formation. Take 5 grams daily, in divided doses.

* *Lecithin* supplements help control cholesterol buildup and prevent gallstones. Lecithin is sold in granule and capsule form at health food stores; follow package directions.

OTHER HELPFUL SUPPLEMENTS

* *Vitamin C* deficiency can cause gallstones. Take 500 to 1,000 milligrams three times daily.

* *Vitamin E* helps prevent fats from becoming rancid. Take 200 to 400 IU daily.

* *Dandelion* enhances liver and gallbladder function. Dandelion is rich in lecithin, which helps control cholesterol. Use commercially available capsules, following package directions.

AN OUNCE OF PREVENTION

To prevent gallstones and gallbladder disease, you should treat your body with the dietary respect it deserves. Eat a diet high in fiber and low in fats, sugars, and cholesterol. Exercise regularly. Avoid smoking or spending time around people who do. Maintain your ideal weight, and avoid crash diets and rapid weight-loss programs, which can cause gallstones.

CALL FOR HELP

Most people with gallstones and gallbladder disease experience the classic symptoms of digestive distress—bloating, gas, and nausea—especially after eating a fatty meal. However, between one-third and one-half of all people with gallstones actually have no pain or warning symptoms—until they suffer from the excruciating pain on the right side of the abdomen or between the shoulder blades, associated with passing a gallstone. If you experience the warning signs, contact your doctor; you should rule out the possibility of appendicitis or another serious illness.

Gingivitis and Periodontal Disease

A beautiful smile requires not only beautiful teeth but healthy gums as well. Periodontal disease (literally meaning "disease around the tooth") is the major cause of tooth loss in older people. To some degree, periodontal disease affects up to 85 percent of the population.

The gums that surround your teeth are called gingival, and the network of gums, bones, and ligaments that form the tooth socket are called the periodontium. When you develop periodontal disease, you can experience swollen, bleeding, and receding gums, as well as loose teeth.

Periodontal disease goes through three stages: gingivitis, periodontitis, and advanced periodontitis or pyorrhea.

- Stage 1 (gingivitis) refers to inflammation of the gums caused by plaque, the sticky bacterial film that forms on the teeth and gums. Plaque continuously builds up on the teeth, where it causes no harm as long as it is removed within twenty-four hours or so. After that time, the plaque hardens into tartar (also known as calculus), which in turn, produces toxins and enzymes that irritate the gums, causing them to become red and swollen. The gums may also bleed during brushing and flossing and begin to recede from the tooth. If treated at this stage, periodontal disease can be controlled, since it has not yet damaged the bone and ligaments that hold the teeth in place.

- Stage 2 (periodontitis) refers to the stage in which the plaque slips beneath the gums and begins to damage the roots of the teeth.

- Stage 3 (advanced periodontitis or pyorrhea) refers to the final stage of the disease, which affects the bones and support system for the teeth. In Stage 3, the gums often recede to the point that the teeth appear elongated; pockets form underneath the gums, where additional plaque and food can collect, causing bad breath and greater gum irritation. The plaque and tartar under the gum line can cause infections that damage the bone, resulting in loose teeth and the loss of teeth.

While inadequate tooth cleaning is the major cause of periodontal disease, other contributing factors include habitual clenching and grinding of the teeth, mouth breathing, a high-sugar diet, and the use of tobacco, drugs, and alcohol. Heredity, hormonal imbalances, and stress are other possible factors.

MOST HELPFUL SUPPLEMENTS

* *Antioxidants*—vitamins A (beta-carotene), C, E and selenium—promote healing of gums. Take up to 10,000 IU beta-carotene daily, 3,000 milligrams of vitamin C daily, 400 IU vitamin E daily, and 400 micrograms selenium daily. In addition, when gums are inflamed, open a capsule of vitamin E oil and rub it directly on the affected area to relieve soreness and promote healing.

* *Coenzyme Q10* relieves symptoms. Take 50 milligrams daily for three weeks.

* *Goldenseal* fights infection when used topically. Mix one teaspoon of goldenseal powder with enough water to form a thick paste, then gently brush your teeth and gums.

* *Quercetin* enhances the potency of vitamin C. Take 300 milligrams three times daily with vitamin C.

OTHER HELPFUL SUPPLEMENTS

* *Chamomile* relieves inflammation. Drink up to 3 cups of the tea daily.

* *Folic acid* reduces inflammation and bleeding. Rinse your mouth with 5 milliliters of a 0.1 percent solution of folic acid twice a day for thirty to sixty days; or take 4 milligrams capsules or tablets daily.

* *Green tea* fights bacteria. Drink up to 3 cups tea daily, preferably with meals. Commercially pre-

pared products are also available; follow package directions.

- *Aloe vera* helps soothe throbbing gums and heal inflamed tissue. Apply the gel directly to the gums before bed.

- *Myrrh* fights infection. Rub the powder on the gums.

AN OUNCE OF PREVENTION

You've heard it before, you'll hear it again: To prevent gum disease and tooth decay, brush and floss your teeth daily. Gum disease usually stems from plaque buildup, so be diligent about brushing your teeth after every meal, if possible. (If you can't brush, at least rinse well with water.) Some dentists believe an electric toothbrush stimulates the gums, in addition to cleaning the teeth, so consider using one if your dentist recommends it. Avoid mouthwashes with alcohol, which can dry out and irritate sensitive gums. And, of course, have your teeth professionally cleaned twice a year, or more often if you already have periodontal disease.

CALL FOR HELP

You may develop periodontal disease without knowing it. Discuss the health of your gums when you visit your dentist for regular checkups. Between dental appointments, look for swollen or bleeding gums, which can also indicate gum problems, and contact your doctor if you notice any abnormalities.

Glaucoma

Glaucoma is characterized by increased pressure within the eye (or intraocular pressure), caused by a buildup of fluid in the eye. The improper flow of fluids appears to be caused by abnormalities in the structure of a protein called collagen in the eye; stress also appears to be a major factor in many cases. Glaucoma affects approximately 2 million Americans and is the major cause of blindness among adults. Nearly 2 percent of people older than forty have glaucoma.

Glaucoma may be acute or chronic. Acute cases usually occur in one eye only and are accompanied by severe throbbing pain in the affected eye, blurred vision, a moderately dilated pupil, nausea, and vomiting. People with chronic glaucoma usually do not have any symptoms during the early part of the disease as the pressure increases slowly but persistently. Loss of vision is gradual and results in tunnel vision.

Treatment of both types of glaucoma involves reducing the amount of pressure in the eye and improving the metabolism of collagen in the eye. Elimination of food allergies is often very helpful in people with chronic glaucoma.

MOST HELPFUL SUPPLEMENTS
- *Ginkgo* reduces intraocular pressure. Take 40 to 80 milligrams three times daily of extract (24 percent ginkgo flavoglycosides).

- *Magnesium* helps reduce intraocular pressure. Take 200 to 600 milligrams daily.

- *Vitamin C* greatly reduces pressure; take as high a dosage as you can tolerate without experiencing diarrhea. Start with 3,000 milligrams in divided doses, and increase the amount you take to 10,000 milligrams, in divided doses (or as tolerated by your digestive system).

OTHER HELPFUL SUPPLEMENTS

- *Bilberry* assists in collagen metabolism. Take 240 to 480 milligrams daily in tablets or capsules standardized to 25 percent anthocyanosides.

- *Chromium* promotes health of eye muscles, especially in diabetics. Take 200 to 400 micrograms daily.

- *Vitamin B complex* helps reduce stress on the nervous system. Take 50 milligrams three times daily with meals.

AN OUNCE OF PREVENTION

Not all cases of glaucoma can be prevented, but you may be able to reduce your risk of developing the disease by eating a diet high in vitamin C and omega-3 fatty acids (salmon, mackerel, herring, and other cold-water fish). Avoid corticosteroid drugs such as prednisone, which weaken collagen structures, including those in the eye.

CALL FOR HELP

Acute glaucoma is a medical emergency; consult your doctor or ophthalmologist immediately. Unless acute glaucoma is adequately treated within 12 to 48 hours, it results in permanent blindness.

Gout

Over the years, gout has earned a reputation as "the rich man's disease" because it tends to afflict overweight, wealthy men who consume diets high in red meat and wine. This description may not be too far from the mark, since eating meat and drinking alcohol can contribute to the disease. Gout affects about

1 million Americans, mostly men over age forty.

Gout is a relatively common type of arthritis caused by the buildup of uric acid, a by-product of the metabolism of purines, which are made by the body and consumed in foods. During a gout attack, the uric acid forms tiny crystals of sodium urate, which collect in a joint (often the big toe), causing inflammation and severe pain. The red-meat-and-wine stereotype makes sense when you consider that meats (especially organ meats) contain high levels of purines, and alcohol (including wine) interferes with the ability of the kidneys to excrete uric acid.

Gout doesn't appear overnight. In some cases, it may take years or even decades for uric acid crystals to build up before an attack. Gout can strike the heel, ankle, or instep, but more than half the time the first attack afflicts the first joint of the big toe. (The big toe may be the most frequent target because uric acid crystallizes more easily at lower temperatures, and the body temperature is somewhat lower in the toes and lower extremities than at the body's core.) Often the acute pain strikes in the middle of the night, especially after overindulging in food and alcohol. Fever and chills may accompany the pain.

About half of all gout sufferers have a second attack within one year, and three-fourths will have a recurrence within four to five years. But chronic gout is rare and unnecessary, since the condition can be controlled by diet and drug therapy designed to lower uric acid levels. Some people develop elevated uric acid levels and gout as a side effect of another disorder, such as kidney disease, which can inhibit uric acid excretion. Low-dose aspirin therapy and some diuretic high-blood-pressure treatments can also cause gout.

MOST HELPFUL SUPPLEMENTS

- *Folic acid* aids in protein metabolism. Studies have found that high levels of folic acid (roughly 10 milligrams per day) inhibit the enzyme responsible for producing uric acid. Caution: Megadoses of folic acid can interfere with drugs used to treat epilepsy, in addition to masking the signs of vitamin B12 deficiency. Only take high doses of folic acid with your doctor's approval.

- *Vitamin B complex* helps to relieve stress, which often accompanies gout attacks. Take 100 milligrams twice daily.

OTHER HELPFUL SUPPLEMENTS

- *Germanium* helps pain and swelling. Take 100 milligrams twice daily.

- *Zinc* aids in protein metabolism. Take 30 to 45 milligrams daily.

- *Nettle* can help treat gout pain when used topically. Though it may sound masochistic, urtication (flailing yourself with the stinging plant) helps relieve gout pain. The fresh plant can also be placed in a juicer to generate nettle juice, which can be applied topically to the affected joint.

- *Molybdenum* is a trace mineral that is a component in several enzymes, including those involved in uric acid formation. Look for a multivitamin–mineral supplement containing molybdenum; additional supplementation is not necessary.

AN OUNCE OF PREVENTION

Lose weight, if necessary. Trimming down to your ideal body weight lowers uric acid levels in the blood, reducing your risk of a gout attack. Also, watch what you eat. Take special care to limit your consumption of foods containing purines, including meat, shellfish, fatty fish, asparagus, mushrooms, spinach, and dried beans. Alcohol should be avoided, because it hinders the elimination of uric acid.

CALL FOR HELP

Doctors can confirm the diagnosis of gout by testing the uric acid levels in the blood, or by withdrawing a sample of joint fluid to look for evidence of crystals. Most cases of gout can be managed by diet alone, but prescription drugs may prove useful as well. Consult your doctor after a gout attack to confirm the diagnosis and to develop a treatment strategy.

Headache and Migraine Headache

When your head hurts, everything hurts. While there are a number of different causes of headaches, there are two main types: tension and migraine headaches. Headache and migraine are covered together because supplement treatments are similar and many of the triggers are the same.

As the name implies, tension headaches usually stem from tension in the muscles of the face, neck, or scalp in response to stress or anxiety. The muscles squeeze the nerves and constrict the blood supply, causing pain and pressure. Tension headaches can be triggered by stress, eyestrain, too much noise or light, grinding of teeth, or poor posture, among other things.

Typically the pain is dull and steady and feels as if there is a band squeezing the head. Often there is tension in the neck and shoulder muscles as well. A sinus headache is caused by congestion within the sinus cavities, which places pressure on the nerves in the face and head. Food allergies and ear infections also can trigger headache pain.

Migraine headaches begin to throb when the blood vessels in the head expand more than normal, often in response to food allergies, hormonal changes, stress, and other factors. Migraines are characterized by pounding, throbbing, sometimes debilitating pain on one side of the head, which may or may not be preceded by auras (visual disturbances), and is usually accompanied by nausea and vomiting. Migraine pain can last for hours or days, and may be accompanied by nausea and vomiting.

MOST HELPFUL SUPPLEMENTS

- *Feverfew* reduces frequency and intensity of headaches. Take ¼ to ½ teaspoon of fluid extract three times daily; take tablets and capsules according to package directions.

- *Ginger* is an antiinflammatory agent that can be quite effective for treating migraine. Take 500-milligram capsules four times daily or take 100 to 200 milligrams of standardized extract three times daily for prevention or every two hours (up to six times daily) to treat acute migraine.

- *5-HTP* raises serotonin levels and helps relieve headache. Chronic headache sufferers tend to have low serotonin levels; several studies have demonstrated excellent results with 5-HTP in the treatment of tension and migraine headaches. Take 100 to 200 milligrams three times daily.

- *Magnesium* levels tend to be low in people who suffer from both tension headache and migraines. There is considerable evidence that low magnesium levels trigger both migraine and tension headaches. Take 250 to 400 milligrams three times daily.

OTHER HELPFUL SUPPLEMENTS

- *Evening primrose oil* helps dilate the blood vessels, easing tension headache pain. Take 500-milligram capsules two to three times daily.

- *Skullcap* helps ease tension headache. Drink up to three cups of tea daily, or use commercial preparations, following package directions. Skullcap is best taken after meals.

- *Valerian* reduces stress and eases headache pain. Make a decoction using 2 teaspoons dried root in 8 ounces of boiling water and simmer for ten minutes. Drink 1 cup daily as needed.

- *SAMe* can help in the prevention of migraine headache, but long-term treatment is required for significant benefits. Use a commercially prepared product; follow package directions. Expect to take SAMe for six weeks before the results become apparent.

AN OUNCE OF PREVENTION

To avoid tension headaches, get regular exercise, talk about your emotional stresses, and learn meditation or other relaxation techniques. If your headaches follow a meal that includes certain trigger foods, you may have a food allergy, and you should avoid the foods that bring on the headaches.

CALL FOR HELP

Most headaches amount to little more than an occasional inconvenience. In rare cases, however, headaches are warning signs of serious health problems, such as meningitis, encephalitis, or brain abnormalities. Call the doctor if the headache follows a head injury, immediately follows a sneeze or sudden cough, occurs each morning along with nausea, or if it is accompanied by fever, stiff neck, lethargy, or vomiting. Also call the doctor if the headaches become more severe over time or become a recurring problem (once a week or more).

Heart Attack and Cardiovascular Disease

Heart problems or cardiovascular disorders are the number-one cause of death in the United States. Heart attacks, congestive heart failure, strokes, and other circulatory disease claim about one million lives a year. In addition, a huge number of Americans—more than 63 million—live with some form of heart or blood vessel disease.

Although the risk of heart attack and cardiovascular disease increases with age, about a fifth of the deaths occur among people under age sixty-five. Fortunately, many of these deaths can be prevented by lifestyle changes and by avoiding or minimizing the factors that raise the risk of cardiovascular disease.

HEART ATTACK

The heart is a muscle, and like any other muscle, it needs oxygen to stay alive. When all or part of the heart muscle dies due to lack of oxygen, it is called a heart attack or myocardial infarction. Each year more than 1.5 million Americans suffer heart attacks, and about one out of three of them dies.

Many heart attacks are caused by blood clots. When blood flows through an artery of the heart that has been narrowed by atherosclerosis (see page 183), it slows down and tends to clot. When the clot becomes big enough, it cuts off the blood supply to the portion of the heart muscle below the clot, and that part of the heart muscle begins to die.

Heart attack can also occur when the heartbeat becomes irregular. In severe cases this condition, known as arrhythmia, can prevent sufficient blood from reaching the heart muscle.

CONGESTIVE HEART FAILURE

When the heart has been damaged and can no longer pump efficiently but has not failed outright, a person is suffering from congestive heart failure. When this occurs, the kidneys respond to the reduced blood circulation by retaining salt and water in the body, which adds additional stress to the heart and makes matters worse.

Congestive heart failure can affect either the right or left side of the heart. The left side pumps oxygen-rich blood from the lungs to the rest of the body. The right side of the heart pumps the oxygen-depleted blood back from the body to the lungs, where the oxygen is replenished. When the left side of the heart is damaged, the blood backs up in the lungs, causing wheezing and shortness of breath (even during rest), fatigue, sleep disturbances, and a dry, hacking nonproductive cough when lying down. When the right side of the heart is damaged, the blood collects in the legs and liver, causing swollen feet and ankles, swollen neck veins, pain below the ribs, fatigue, and lethargy.

STROKE

A stroke is like a heart attack in the brain. Just as a part of the heart dies when deprived of oxygen during a heart attack, so a part of the brain dies when deprived of oxygen during a stroke. A *thrombic stroke* occurs when an artery in the brain is blocked by a clot or atherosclerosis; an *embolic stroke* occurs when a small clot (known as an embolus) forms elsewhere in the body and moves to the brain, where it lodges in an artery and blocks the flow of blood. A *hemorrhagic stroke* occurs when an artery ruptures, usually due to high blood pressure. While hemorrhagic strokes are less common—only about 20 percent of all strokes—they are much more lethal, causing about 50 percent of all stroke-related deaths.

In the aftermath of a stroke, the person loses the bodily functions association with the part of the brain that was destroyed. Symptoms of a stroke include slurred speech or loss of speech, sudden severe headache, double vision or blindness, sudden weakness or loss of sensation in the limbs, or loss of consciousness. These symptoms can occur over a period of a few minutes or hours, and they can occur on one side of the body or both.

Stroke is the nation's third leading cause of death and the leading cause of adult disability. Experts estimate that as many as 80 percent of all strokes can be prevented, either through changes in lifestyle or through the use of drugs to control high blood pressure and the tendency to form blood clots.

MOST HELPFUL SUPPLEMENTS

- *Calcium* aids function of the heart muscle. Calcium is essential for blood clotting; it also plays a role in maintaining blood pressure. Take 1,500 milligrams daily, after meals and at bedtime.

- *Carnitine* reduces fat and triglyceride levels in the blood. Several double-blind studies have shown that carnitine improves heart function and improves symptoms in people with congestive heart failure. Take 500 milligrams two to three times daily on an empty stomach.

- *Coenzyme Q10* oxygenizes the blood and improves exercise tolerance. Take 50 to 100 milligrams three times daily.

- *Garlic* improves circulation. This herb contains several sulfur compounds that block the biosynthesis of cholesterol. Garlic also helps expand the blood vessel walls, increasing blood flow and lowering blood pressure. Another chemical in garlic, ajoene, helps prevent blood clots. Use garlic liberally in cooking, eat up to 6 cloves of fresh garlic daily, or use commercially prepared odorless garlic capsules.

- *Folic acid* and other B vitamins play a key role in recycling homocysteine, a by-product of protein breakdown that has been linked with increased risk of heart attack and stroke. Without enough folic acid and B vitamins, the recycling of homocysteine into methionine becomes inefficient, and homocysteine levels increase. Increased intake of folic acid, vitamin B6, and vitamin B12 decreases homocysteine levels. Take 400 milligrams daily.

- *B vitamin complex* helps decrease homocysteine levels. Take one dose daily.

OTHER HELPFUL SUPPLEMENTS

- *Magnesium* reduces the risk of arrhythmias and replaces magnesium depleted by drugs given for congestive heart failure. Magnesium is necessary to activate an enzyme that helps transport potassium to the cells. Take 750 milligrams daily.

- *Lecithin* emulsifies fat in the blood. Take 2 capsules or 1 tablespoon with meals.

- *Evening primrose oil* helps prevent hardening of the arteries. Take two 500-milligram capsules three times daily.

- *Green tea* lowers blood pressure and cholesterol. Drink up to 5 cups of tea daily, preferably with meals, or take 500-milligram capsules daily.

- *Hawthorn* increases blood flow to the heart, strengthens heart contractions. This "heart herb" enhances cardiac output, in addition to opening up the peripheral vessels to improve overall circulation. Use a commercial product, follow package directions.

- *Vitamin E* strengthens the immune system and heart muscle. Take 400 IU daily.

- *Ginkgo* helps dilate the blood vessels and improves

overall circulation. Use a commercial product; follow package directions.

- *Selenium* is a potent antioxidant useful in the prevention of cardiovascular disease. Take up to 200 micrograms daily.

AN OUNCE OF PREVENTION

The major risk factors for heart disease and circulatory problems include smoking, high blood pressure, obesity, high blood cholesterol levels, and a family history of the disease. There are steps you can take to reduce many of these risks.

- Stop smoking. Smoking constricts the arteries, raises blood pressure, increases arterial tearing, speeds atherosclerosis, and reduces oxygen levels in the blood. Smokers have two to four times the risk of heart attack as nonsmokers, and their heart attacks are more likely to be fatal. But there is hope: A decade after quitting, a former pack-a-day smoker has almost the same heart attack risk as if he or she had never smoked.

- Exercise to keep your arteries strong and flexible. Aerobic exercise helps prevent cardiovascular disease by lowering LDL cholesterol levels and raising HDL cholesterol levels, reducing blood pressure, keeping weight down, burning fat, lowering blood-sugar levels, and boosting relaxation. People who exercise regularly are about half as likely as sedentary people to have a heart attack.

- Maintain a healthy body weight. Excess body fat increases blood pressure and adds stress to the heart and circulatory system. People who maintain their ideal body weight are 35 to 55 percent less likely to have a heart attack than those who are obese.

- Become more aware of your anger, anxiety, and fear. These negative emotions trigger the release of adrenaline and increase blood pressure. These hormones also encourage the cells to release fat and cholesterol into the bloodstream.

- Review all prescription and over-the-counter medicines with your doctor.

- Lower your blood cholesterol, if necessary.

- Keep your blood pressure out of the danger zone (see page 266).

- Know your family history. Anyone whose parents or other close relatives have suffered a heart attack or stroke before age fifty-five should make a special effort to minimize their other risk factors.

CALL FOR HELP

Heart attack victims often delay seeking medical help, frequently with fatal results. Most heart attack deaths occur in the first two hours, yet studies have found that many people wait four to six hours to get to an emergency room. Never ignore the warning signs of a heart attack, which include the following.

- chest pain: an uncomfortable pressure, fullness, squeezing, or crushing feeling in the center of the chest that lasts two minutes or longer

- severe pain that radiates to the shoulders, neck, arms, jaw, or top of the stomach

- shortness of breath

- paleness

- sweating

- rapid or irregular pulse

- dizziness, fainting, or loss of consciousness

Not all of these warning signs occur in every heart attack. And some people, especially older people and diabetics, may not experience symptoms during a heart attack. (These so-called "silent heart attacks" can be detected only by an electrocardiogram.) If you suspect you may be experiencing a heart attack, get emergency help immediately.

Heartburn and Indigestion

Heartburn is a burning sensation in the chest behind the breastbone that sometimes happens after eating. The condition—which may last for several minutes to several hours—occurs when the muscular opening from the esophagus into the stomach doesn't work properly, allowing hydrochloric acid to wash back up into the throat. The hydrochloric acid, which is used by the stomach during digestion, is produced in abundance when you consume spicy, fried, or fatty foods, as well as coffee, citrus fruits, chocolate, tomatoes or tomato-based foods, and alcohol. Heartburn may also be the result of ulcers, gallbladder problems, allergies, stress, hiatial hernia, or an enzyme deficiency.

MOST HELPFUL SUPPLEMENTS

- *Ginger* eases heartburn. Make a decoction by placing 2 teaspoons dried gingerroot in 8 ounces of boiling water and let steep ten minutes. Drink up to 2 cups per day.

- *Licorice* eases symptoms of heartburn. Take one 200-to-300-milligram DGL chewable tablet three times daily before meals.

- *Peppermint* eases heartburn. This herb is part of a class of herbs that relaxes the muscles involved in the digestive process. Drink up to 3 cups infusion daily, or take 2 to 3 capsules daily between meals.

OTHER HELPFUL SUPPLEMENTS

- *Sulfur* relieves symptoms of heartburn. Take 3,000 milligrams daily for one week.

AN OUNCE OF PREVENTION

To minimize your risk of developing heartburn, avoid trigger foods (listed above), avoid lying down after eating, and avoid the use of over-the-counter antacids, especially those that contain aluminum and sodium. Eat small, frequent meals; avoid overeating; avoid drinking lots of liquids with your meals.

CALL FOR HELP

Contact your doctor if heartburn is accompanied by shortness of breath, dizziness, vomiting, diarrhea, severe abdominal pain, fever, sweating, or blood in the stool. In combination with these symptoms, heartburn may signal a more serious condition, such as ulcers, hernia, gastritis, or heart attack.

Hemorrhoids

Most people keep it a secret, but four out of five Americans experience hemorrhoids at some point in their lives. Hemorrhoids, those inflamed and widened veins that look like purple skin growth around the anus, can be painful, itchy, and embarrassing.

Hemorrhoids—also known as piles—can appear either inside or outside the anus. They can be caused by constipation, pregnancy, poor diet, lack of exercise, heavy lifting, obesity, allergies, prolonged sitting, and liver damage. The condition tends to run in families, either because people inherit delicate veins or because they acquire similar lifestyles and personal habits that can exacerbate the problem. Either way, many cases of hemorrhoids can be prevented by carefully monitoring bowel function.

MOST HELPFUL SUPPLEMENTS

- *Calcium* helps blood clotting and prevents cancer of the colon. Take 1,500 milligrams daily. Take with 750 milligrams magnesium.

- *Bilberry* has been shown in a number of studies to strengthen vein walls. It contains powerful anthocyanosides, which improve circulation. Take 100 milligrams of standardized extract three times daily.

- *Psyllium* helps keep the colon clear and relieves pressure in the rectum. Take 7 grams three times daily.

- *Vitamin C* promotes healing and blood clotting. Take 3,000 to 5,000 milligrams daily.

• *Vitamin E* promotes healing and blood clotting. Take 600 IU daily.

AN OUNCE OF PREVENTION

While some unfortunate people tend to be more susceptible to hemorrhoids than others, you can avoid many episodes by not straining or pushing during bowel movements. You can also prevent hemorrhoids in many cases by following the same regimen used to treat them—by getting regular exercise, drinking at least eight 8-ounce glasses of water a day, and eating a high-fiber diet rich in fresh fruits and vegetables.

CALL FOR HELP

If you develop hemorrhoids, you'll know it. Symptoms include rectal pain and tenderness, itching, and sometimes bleeding. In many cases the first—and most startling—sign of hemorrhoids is bright red blood in the stool, caused by pressure during the bowel movement. The bleeding is usually minor. But dark, tarry blood could include a more severe internal bleeding problem and should be brought to a medical doctor's attention immediately.

Herpes

Herpes is an extremely common—and contagious—viral disease that causes painful cold sores and genital lesions. The herpes virus has been classified into two types: Type I (herpes simplex-1 virus), which causes skin eruptions and cold sores around the mouth; and Type II, also known as genital herpes, which is transmitted sexually. Type II (herpes simplex-2 virus) is the more prevalent infection. Once the virus enters the body, it remains throughout the individual's life and periodically becomes active and symptomatic. About 30 percent of the population suffers from periodic herpes outbreaks.

Among women, the symptoms include the appearance of blisters around the clitoris, cervix, rectum, and vagina, as well as a watery discharge from the urethra. Some women experience pain when urinating. Men experience blisters on the penis, scrotum, and groin, swollen lymph nodes in the groin, and pus when urinating. In both men and women, the blisters erupt within several days of their appearance, leaving behind large, painful ulcers. The ulcers eventually dry up and heal.

In some people the infection becomes severe and causes serious brain damage and inflammation of the liver. Babies can contract the disease at birth, which can result in brain damage, blindness, and death.

Stress and immune system weakness can cause repeat outbreaks after the initial blisters clear up. In addition, eating foods high in the amino acid arginine stimulates viral activity; arginine is found in chocolate, peanut butter, nuts, and seeds. Other foods believed to increase the chance of an outbreak include alcohol, sugar, refined carbohydrates, coffee, and processed foods.

MOST HELPFUL SUPPPLEMENTS

- *Lysine* has been used to treat cold sores and herpes Type I for years. The amino acid prevents the herpes virus from reproducing. Take 500 milligrams daily.

- *Echinacea* is a powerful immune system stimulant that can help heal herpes outbreaks. Use commercially prepared products, follow package directions.

- *Vitamin A* promotes healing and prevents spread of infection. Take 50,000 IU daily.

- *Vitamin B complex* helps relieve stress. Take 50 milligrams up to three times daily.

- *Zinc* lozenges can help relieve sores in and around the mouth. Take 50 to 100 milligrams daily in divided doses, or suck zinc lozenges, following package directions. You may also use topical zinc sulfate ointment to heal sores.

OTHER HELPFUL SUPPLEMENTS
- *Acidophilus* prevents growth of harmful bacteria. Take 3 capsules daily on an empty stomach.

- *Garlic* is a natural antibiotic. Take 3 tablets three times daily with meals.

- *Goldenseal* helps fight viral infection. Take 2 capsules three times daily.

- *Licorice* inactivates the herpes virus when used topically. Apply the extract glycyrrhizin three to four times daily to the affected area.

AN OUNCE OF PREVENTION
Herpes is forever; once you have been infected with the virus, you will have to live with the possibility of outbreak for the rest of your life. The only way to prevent the disease is to avoid unprotected sexual contact with an infected partner.

Once you have the virus, you can minimize your risk of subsequent outbreaks by maintaining a strong immune system and reducing stress. In addition, the amino acid lysine appears to help keep the herpes virus at bay;

good food sources of lysine include kidney beans, lima beans, soybeans, split peas, corn, and potatoes.

CALL FOR HELP

If you experience the symptoms of herpes, contact your doctor for an accurate diagnosis and treatment plan.

Hypertension (High Blood Pressure)

Hypertension is more than a medical annoyance; it is the most accurate predictor of future cardiovascular disease in people over age sixty-five. Simply put, hypertension refers to the pressure of the blood against the blood vessels as the heart pumps through the arteries. When blood presses against the walls of the blood vessels at a level that is greater than normal, it is known as hypertension, or high blood pressure.

One of the main reasons the blood does this is that the arteries are clogged with cholesterol, forcing the heart to work harder and harder to pump the blood through narrow vessels, which in turn raises blood pressure. Other factors that can lead or contribute to hypertension include smoking, obesity, high salt intake, use of contraceptives, stress, and excessive intake of coffee or tea.

A blood pressure reading consists of two numbers: the systolic pressure (the higher number, reflecting the pressure when the heart contracts) and the diastolic pressure (the lower number, reflecting the pressure as the heart rests between beats). A normal blood pressure reading is defined as 120/80, though the numbers fluctuate somewhat throughout the day. Blood pressure is measured with an instrument called a sphygmomanometer, and the numbers refer to the level to which a column of mercury (Hg) rises at each pressure reading.

Approximately 40 million Americans have high blood pressure, which places them at risk for heart failure, stroke, loss of vision, and kidney failure because the heart must work harder than normal to pump blood. High blood pressure is often seen in people who have diabetes, coronary heart disease, kidney problems, obesity, arteriosclerosis, and adrenal tumors. Many people with hypertension are unaware they have the condition because they experience no symptoms.

Fortunately, in most cases high blood pressure can be lowered by diet and lifestyle changes. Hypertension should not be considered an inevitable consequence of aging; it can be controlled.

MOST HELPFUL SUPPLEMENTS

- *Calcium* deficiency is common in people with hypertension. Several clinical studies have demonstrated that calcium can lower blood pressure, especially among African Americans, people who are salt sensitive, pregnant women, and the elderly. Typical reductions in blood pressure reported in these groups are about 10 mm Hg (systolic) and 5 mm Hg (diastolic). Take 1,500 milligrams daily.

- *Coenzyme Q10* reduces blood pressure and improves heart function. Take 100 milligrams daily.

- *Garlic* is a powerful blood-pressure-lowering agent; it has been shown to lower the systolic pressure by an impressive 20 to 30 mm Hg and the diastolic by 10 to 20 mm Hg. Consume three to ten cloves of fresh garlic daily, or use commercial garlic preparations, following package directions.

OTHER HELPFUL SUPPLEMENTS

- *Carnitine* helps prevent heart disease. Take 500 milligrams twice daily on an empty stomach.

- *Hawthorn*, widely used in Europe, lowers blood pressure, improves coronary circulation, prevents cholesterol deposits from forming on artery walls, and helps strengthen the heart muscle. Use a commercial product; follow package directions.

- *Potassium* deficiencies are possible in people taking diuretics for blood-pressure control. Low levels of potassium, especially when combined with high levels of sodium, increase fluid retention and inhibit the body's blood-pressure-regulating system. If you are taking diuretics or high-blood-pressure medication, eat a banana every other day or take 99 milligrams daily.

- *Vitamin C* has been found to help lower blood pressure. One study found that taking as little as 250 milligrams of vitamin C daily slashed the risk of high blood pressure by almost half. Take 1,500 milligrams of vitamin C daily.

AN OUNCE OF PREVENTION

Not every case of hypertension can be controlled or prevented by lifestyle changes, but many can be. Keep the following tips in mind.

- Know your blood pressure.

- Don't smoke, and avoid being around people who do. Nicotine constricts the arteries and elevates blood pressure. A person with uncontrolled hypertension who smokes is five times more likely to have a heart attack and sixteen times more likely to have a stroke than a nonsmoker.

- Exercise regularly.

- Lose weight, if necessary. People who are overweight experience more hypertension. An analysis of five studies involving weight loss and hypertension found that, on average, losing 20 pounds resulted in a decline of 6.3 mm Hg in systolic and 3.1 mm Hg in diastolic pressure.

- Monitor your use of over-the-counter medicines. Avoid using antihistamines, decongestants, cold remedies, and appetite suppressants, unless recommended by a doctor.

- Try to manage your daily stress. Stress can temporarily elevate blood pressure. Relaxation techniques such as biofeedback, meditation, yoga, progressive muscle relaxation, and hypnosis have been shown to help lower blood pressure.

- Avoid coffee and caffeinated beverages.

- Restrict alcohol consumption. While some researchers tout the cardiovascular benefits of modest drinking, consuming more than 30 milliliters of alcohol a day—an amount equal to 1 ounce of 100-proof whiskey, 8 ounces of wine, or two 12-ounce beers—can raise blood pressure.

CALL FOR HELP

Hypertension is often called "the silent killer" because it strikes without warning. About 20 percent of Americans with high blood pressure don't know they have the condition, and only one-third have it under control. Advanced hypertension can cause headache (especially in the morning), fatigue, dizziness, rapid pulse, shortness of breath, nosebleeds, and visual problems. If you experience these symptoms, contact your doctor promptly.

The only way to be sure your blood pressure is under control is to visit your doctor regularly and have your blood pressure checked. If you suffer from hypertension, you need to be under the care of a physician who can monitor this potentially life-threatening condition.

Impotence (Erectile Dysfunction)

Sooner or later, it happens to almost *every* man. Still, most men don't like to talk about impotence, a chronic problem in achieving and maintaining an erection long enough to experience intercourse. But impotence is nothing to feel ashamed of or humiliated about.

As an inevitable—and normal—part of the aging process, the speed of a man's sexual response slows and the intensity of orgasm declines in response to a drop in testosterone levels, as well as to a decrease in blood circulation. In addition to reducing the need or desire to reach orgasm, this drop in hormone levels can result in the production of a smaller amount of semen during ejaculation. Don't think of these changes—which usually begin to occur in the late forties—as problems. Consider the benefit: You may find that you can enjoy intercourse longer before ejaculation.

And even if you do experience impotence rather than a gradual decline of sexual function associated with aging, you can rest assured that you are not alone. It is estimated that more than 10 to 20 million American men are chronically impotent, including a quarter of men over age 65 and more than half of all men over 75.

Roughly three-fourths of all erection problems have at least some physical cause. To achieve an erection, there must be cooperation among blood vessels, nerves, and tissues. A number of health problems—including diabetes, heart, and circulation problems, stroke, epilepsy, Alzheimer's disease, neurological disorders, alcohol and drug abuse, Parkinson's disease, and liver and kidney disease—can cause impotence. So can certain medications, including tranquilizers, diuretics, and antiulcer, antipsychotic, antidepressant, and antihypertensive drugs. Some over-the-counter antihistamines and decongestants can cause temporary impotence.

Other cases of impotence stem from psychological factors, such as relationship problems, stress, anxiety, grief, depression, fatigue, boredom, and guilt. Sexual intimacy can make some people feel very vulnerable, causing a number of stresses and uncomfortable feelings. With patience and treatment, most cases of psychological, as well as physical, impotence can be managed and overcome.

MOST HELPFUL SUPPLEMENTS
- *Ginkgo* helps improve blood flow to the penis, which can help combat impotence. Ginkgo is usually available only in commercial preparations; follow package directions.

- *Yohimbe* is the primary ingredient in a prescription drug used to treat impotence by increasing blood flow. It is commercially available; follow package directions. It is often included in "male potency formulas" and other supplements designed for men.

OTHER HELPFUL SUPPLEMENTS

- *Ginseng* has long been considered a mild aphrodisiac. Ginseng preparations are commercially available; follow package directions.

- *Vitamin A* is necessary for the body to produce the sex hormones that are essential for sexual function. Take a multivitamin–mineral supplement, which should provide an adequate amount of this important vitamin.

- *Zinc* assists with prostate gland function and overall reproductive health. Take 15 to 50 milligrams daily.

AN OUNCE OF PREVENTION

Being in good physical condition can improve potency as well as overall health. Eat a well-balanced diet, and get regular exercise. Avoid alcohol, which can cause temporary impotence, as well as smoking, which restricts blood flow throughout the body. When making love, take your time, relax, and enjoy some foreplay. Have intercourse more often, since testosterone levels remain higher in men who have sex more frequently.

CALL FOR HELP

Impotence shows up at the least convenient times—and with no warning or invitation. A single episode should not be cause for alarm, but a pattern of difficulty maintaining an erection merits a discussion with a medical professional to rule out physical problems.

If natural remedies don't help and your doctor can't find a physical basis for the problem, contact a psychologist or mental health professional. Statistics indicate that therapy can help in four out of five cases of psychologically based impotence.

Incontinence

Urinary incontinence (or loss of bladder control) can be embarrassing—even humiliating—but many people understand the problem because they experience it themselves. More than 10 million people, including at least 10 to 20 percent of all older adults—live with incontinence, though most won't talk about it much.

Before age sixty-five, incontinence affects three to five times more women than men. Women who have had children experience more problems because pregnancy places intense pressure on the bladder and muscles of the pelvic floor; in addition, labor and delivery can tear the muscles and surrounding tissues, sometimes leaving them less resilient than before. About 40 percent of women experience some incontinence during pregnancy, and 10 percent continue to have problems afterward. At menopause, the decrease in estrogen can weaken the pelvic floor muscles and thin the lining of the urethra, loosening the seal at the neck of the bladder.

Men experience less incontinence, in part because they have longer urethras (10 inches, versus about 2 inches for women). The prostate gland also helps support a man's urethra, helping to prevent leakage. An enlarged prostate, however, can put pressure on the bladder, so that after age sixty-five men and women have an almost equal chance of becoming incontinent.

Still, you don't have to accept incontinence as an inevitable part of aging. Most cases of incontinence can be either controlled or cured. Incontinence isn't a disease, but a symptom of an underlying problem, such as weak muscles in the pelvic floor or an obstruction of the bladder outflow.

There are five basic types of chronic incontinence.

- Stress incontinence is a condition in which small amounts of urine dribble out when you exercise, cough, laugh, sneeze, or move in other ways that put pressure on the bladder. Most cases of stress incontinence are associated with weak muscles in the pelvic floor, though in severe cases there may be nerve damage or tears in the sphincter muscles.

- Urge incontinence usually involves the loss of large amounts of urine with little warning. It occurs when the need to urinate comes on so quickly that there isn't enough time to make it to the toilet. Urge incontinence can be caused by stroke, Parkinson's disease, kidney or bladder stones, or bladder infection.

- Overflow incontinence involves urination with no warning or sensation. In such cases the urine overflows and spills out when a person shifts position or stands up. Often the person feels the need to urinate again a few minutes later but cannot empty the bladder completely. People with overflow incontinence have a high risk of bladder infection. It sometimes occurs following pelvic surgery or a bladder suspension operation. Diabetes or an enlarged prostate can contribute to overflow incontinence as well.

- Reflex incontinence involves involuntary, spontaneous urination—no warnings, no urges, no rush to the bathroom. This lack of bladder control is usually caused by spinal cord injury, diabetes, multiple sclerosis, and other serious medical conditions.

- Functional incontinence strikes people who have normal bladder control and warnings but cannot reach the bathroom fast enough due to physical limitations.

In addition, temporary incontinence can be caused by the use of diuretics and other medications. Most cases of incontinence can be controlled or managed by correcting the underlying health problem or condition.

MOST HELPFUL SUPPLEMENTS
- *Black cohosh* has mild estrogen effects, meaning it acts like the female hormone estrogen, which can help strengthen the muscles in the pelvic floor. Commercial products are available, follow package directions.

AN OUNCE OF PREVENTION
Many people with incontinence simply have lost the muscle tone in their pelvic floor, which supports the bladder, uterus, and other internal organs. The easiest way to deal with this problem is to strengthen the muscles by doing the Kegel exercise. To tone and strength the pelvic floor, follow these steps.

- Locate the appropriate muscles by repeatedly stopping your urine in midstream. The muscles you squeeze around your urethra and anus to stop the urine are the muscles you want to work on.

- Practice squeezing, then releasing these muscles several times each day when you urinate. Once you are familiar with them, practice squeezing them when you are not urinating.

- During each exercise contraction, hold the squeeze for three seconds, then relax for three seconds. Repeat this 10 to 15 times per session, three or four sessions a day.

Extra pounds can also cause the pelvic floor to sag, giving you yet another good reason to shed the extra weight. It's also a reason to quit smoking; nicotine in cigarettes irritates the bladder and a smoker's cough can cause problems with stress incontinence.

While waiting for these lifestyle changes to take effect, try the technique known as "double voiding." Empty your bladder, then relax a minute and try again. You might also try urinating, then standing up for a minute, sitting down and leaning forward, then trying again.

CALL FOR HELP
If you experience incontinence, talk to your doctor about the problem. More than half of all people with incontinence fail to seek help, though experts estimate that more than 80 percent can overcome the problem.

Infertility

Technically speaking, infertility is the inability to become pregnant after having unprotected sexual intercourse regularly during ovulation for a period of twelve months or more, or the inability to carry a pregnancy to full term. Every couple that has tried unsuccessfully to become pregnant knows that infertility involves much more than that—it entails a monthly roller-coaster ride of promise and possibility, followed by crashing heartbreak.

Responsibility for infertility is evenly divided between men and women: about 40 percent of cases can be traced to a problem with the woman, 40 percent can be traced to the man, and 20 percent involve both partners.

The most common causes of infertility among women are a failure to ovulate (which is often caused by a hormone imbalance) and a blocked passage of the egg from the ovary to the uterus (which is often associated with endometriosis, infection, or growths). Other causes include sexually transmitted diseases, pelvic inflammatory disease, smoking, excessive consumption of caffeine, being overweight or underweight, and age (fertility decreases after age thirty-five). Occasionally an iron deficiency causes infertility in women. Before taking an iron supplement, however, an iron deficiency should be verified by a physician.

Among men the problem may be a low sperm count, low motility (sperm movement is impaired), malformed sperm, or blocked sperm ducts. Wearing tight underwear or pants may temporarily raise the temperature of the testicles, which reduces sperm production. In both men and women fertility can be adversely affected by depression, anxiety, or exposure to radiation, pesticides, or other environmental poisons.

MOST HELPFUL SUPPLEMENTS
- *Chaste berry* may be helpful in women who have high levels of the hormone prolactin and a shortened second half of the menstrual cycle. Take 1 capsule daily, following package directions.

- *Arginine* helps improve male sperm count. In one study, 74 percent of 178 men with low sperm count had significant improvements in sperm count and motility after taking arginine. Commercial products are available; follow package directions.

- *Vitamin E* helps balance hormone production in both men and women. Take 400 to 1,000 IU daily.

- *Zinc* improves testosterone levels and sperm counts in men. Zinc is involved in hormone metabolism, sperm formation, and sperm motility. Take 25 milligrams three times daily.

OTHER HELPFUL SUPPLEMENTS

- *Carnitine* improves sperm count or motility. Take 3 grams daily for four months.

- *Vitamin C* plays an important role in sperm formation. There is much more vitamin C in seminal fluid than in other body fluids, including blood. Several double-blind studies have shown that vitamin C increases sperm count. In one study, one week after thirty men received 1,000 milligrams of vitamin C daily, they demonstrated a 140 percent increase in sperm count. Take 1,500 milligrams daily, in divided doses.

AN OUNCE OF PREVENTION

Several months before you plan to begin trying to conceive, take steps to minimize your consumption of alcohol and medications, including over-the-counter drugs. Men should avoid hot tubs, which can impair sperm production.

CALL FOR HELP

You should talk to your doctor about your plans to start a family before you begin trying to conceive. Your doctor may want to do some blood tests to ensure that you are in optimal health before you plan your pregnancy. If you have been trying to conceive for six months or more and are thirty-five or older, contact your doctor. Your doctor will probably want to test your hormone levels and run a few screening tests to see if you have a simple problem that is affecting your fertility, or you may need a referral to a fertility expert.

Insomnia

Insomnia can be a nightmare. You're desperate for sleep, but it seems that the more exhausted you feel, the harder it is to rest.

Up to 30 percent of Americans have insomnia, which refers to any of three sleep disorders: difficulty falling asleep (more than forty-five minutes), early morning awakening, or frequent night awakenings (six or more a night). You have insomnia, however, only if you experience these symptoms and they leave you tired and worn down. After all, you are the ultimate judge of how much pillow time your body needs.

Approximately ten million people take prescription drugs to help them sleep, while many more take over-the-counter sleep aids. One of the major problems with taking drugs to help you sleep is that long-term or chronic use of the sleeping pills causes addiction (in the case of drugs in the benzodiazepine class) and disturbing side effects, including abnormal sleep patterns, which actually make the insomnia worse.

Studies in sleep laboratories show that 50 percent of all cases of insomnia are caused by psychological factors, especially depression, anxiety, and tension. Other causes include use of drugs (including caffeine and alcohol), hypoglycemia (low blood sugar), changes in a person's environment, pain, restless leg syndrome (an uncontrollable urge to move the legs), and fear of sleep. Natural approaches to eliminating insomnia include getting enough exercise, avoiding foods, beverages, and drugs that contain caffeine, and practicing relaxation techniques.

MOST HELPFUL SUPPLEMENTS

- *5-HTP* induces sleep. Take 100 milligrams daily.

- *Skullcap* causes drowsiness. Take 1 teaspoon in warm water before retiring.

- *Valerian* contains chemicals known as valepotriates, which have sedative properties. Take any one of the following forty-five minutes before retiring: 2 to 3 grams dried root in 8 ounces of boiling water as a decoction; 4 to 6 milliliters tincture; 1 to 2 milliliters fluid extract; 150 to 300 milligrams dry powdered extract.

- *Calcium and magnesium* deficiency can cause you to wake up after a few hours of sleep and have trouble going back to sleep. Take 1,500 milligrams of calcium and 750 milligrams of magnesium daily.

OTHER HELPFUL SUPPLEMENTS

- *Vitamin B6* (pyridoxine) helps reduce stress. Take 50 milligrams forty-five minutes before retiring.

- *Melatonin* is secreted by the pineal gland in the brain to help regulate the body's sleep-wake cycle. Melatonin levels drop as we age, so a supplement may help induce sleep. Supplements are commercially available; follow package directions.

AN OUNCE OF PREVENTION

To prevent insomnia, get regular exercise, but avoid strenuous exercise within two hours before bedtime Avoid smoking or consuming caffeine or alcohol for three or four hours before bedtime. That before-bed nightcap might make you feel sleepy at first, but it will make you more likely to awaken during the right. Sex is a natural relaxant; under the right circumstances, intercourse can help you unwind and enjoy a good night's sleep.

If you need to awaken in the middle of the night, your blood sugar levels may be falling too low. To avoid the problem, try having a high-carbohydrate snack before bed. Milk, whole-grain crackers, peanut butter, and bananas are particularly good choices because they contain the sleep-inducing chemical tryptophan, in addition to being good sources of carbohydrates, which can help your body maintain moderate blood sugar levels into the night.

Fortunately, it's relatively easy for the body to "catch up" on sleep. A single night of full sleep—one in which you sleep until you naturally awaken—will allow you to regain about 90 percent of the mental sharpness you lost due to sleep deprivation. Add a second full night and you should be as sharp as ever.

CALL FOR HELP

If you have insomnia, you'll recognize it by the sleepy feeling during the day and the restless feeling at night. In addition to sleepless nights, you will almost certainly experience trouble concentrating and daytime fatigue and irritability. Consult your doctor if your insomnia lingers for more than a few weeks, or if the condition begins to interfere with your feelings of physical or mental stability. Your doctor should review all the prescription and over-the-counter medications you're taking to find out if they could be contributing to the problem. If the doctor suspects a physical cause of your insomnia, you may be referred to a sleep specialist, who will monitor your sleep patterns and brain-wave patterns during the night. If psychological stress contributes to the problem, you may be referred to a psychologist or psychiatrist to explore other issues related to the problem.

Your doctor may also want to rule out sleep apnea, a common problem in which sleepers stop breathing for periods of ten seconds or more in the night. Left untreated, apnea can cause serious health problems, such as high blood pressure, an enlarged heart, and stroke.

Irritable Bowel Syndrome and Inflammatory Bowel Disease

Irritable bowel syndrome (IBS) and inflammatory bowel disease refer to a range of digestive complaints that can cause diarrhea, constipation, bloating, and cramping. While many of the natural treatments are shared, these conditions vary in their symptoms and severity.

Irritable bowel syndrome is a very common disorder

that affects the large intestine and causes it to have irregular muscular contractions. IBS is a functional disorder of the intestine, meaning it does not involve any
kind of structural defect. This abnormal action prevents waste materials from moving efficiently through
the intestines, leading to a buildup of toxins and causing bloating, gas, abdominal pain, and alternating diarrhea and constipation. People with irritable bowel
syndrome often experience back pain and fatigue.
Symptoms of the syndrome worsen in some women
before and during their menstrual cycle.

Researchers have not determined the exact cause of
irritable bowel syndrome, although food allergies and
stress both appear to play a major role. Foods to avoid
include meat and dairy products, spicy foods, fried
foods, processed foods, sugar, coffee, alcohol, and soft
drinks. Wheat and wheat products cause symptoms in
some individuals.

Inflammatory bowel disease (IBD) is a broad term
for several chronic inflammatory disorders that affect
the intestinal tract; Crohn's disease and ulcerative colitis are the two main categories of IBD. All forms of
IBD share the characteristic signs of recurrent inflammation of various portions of the intestines, although
symptoms may differ. Causes for these diseases are
unknown, although food allergies, heredity, infection,
and antibiotic use appear to play a role.

Crohn's disease can affect the entire digestive system, including the small and large intestines, stomach,
esophagus, and mouth. The disease typically first
strikes around age twenty and can recur every few
months or stay dormant for years. Symptoms include
diarrhea, cramps, lower right abdominal pain, loss of
energy, weight loss, lack of appetite, fever, and malnu-

trition. Recurrent attacks weaken the intestines and can cause bowel function to worsen.

Ulcerative colitis is present when the lining of the colon is inflamed and marked with ulcers. The disease is usually limited to the colon. Symptoms include bloody diarrhea, cramps, fever, frequent need to defecate, and abdominal tenderness.

People with IBD often have deficiencies of iron, vitamin B12, folic acid, magnesium, potassium, vitamin D, and zinc; less often low levels of vitamin K, copper, niacin, and vitamin E are seen. Correction of these deficiencies is important, especially in children with IBD, because many of them fail to grow properly. A high-potency multivitamin–mineral supplement is recommended as part of any treatment program.

MOST HELPFUL SUPPLEMENTS

- *Acidophilus* helps digestion and helps to maintain a healthy bacterial balance in the digestive tract. Take 2 to 3 billion live organisms daily (about 1 teaspoon powder or liquid in water twice daily).

- *Peppermint* tea and peppermint oil have long been used to calm intestinal spasms. To be most effective, peppermint oil capsules should be enteric-coated to prevent the oil from being released in the stomach. With the coating, the peppermint oil travels to the small and large intestines, where it relaxes the intestinal muscles and promotes the elimination of excess gas. Commercial products are available; follow package directions.

- *Vitamin C* and *vitamin E* are antioxidants that prevent damage to the intestinal wall and promote tissue healing. Take 1,500 milligrams of vitamin C and 200 to 400 IU of vitamin E daily.

* *Zinc* can be used to help repair intestinal walls. Take 25 to 50 milligrams daily.

OTHER HELPFUL SUPPLEMENTS

* *Omega-3 fatty acids* help to reduce inflammation. Commercial products are available; follow package directions.

* *Vitamin B complex* restores possible nutritional deficits and helps to repair intestinal walls. Take one dose daily.

* *Alfalfa* improves digestion and cleanses the blood. Take 1 tablet or 1 tablespoon liquid three times daily.

* *Chamomile* soothes and tones the digestive tract and helps ease alternating diarrhea and constipation. Drink chamomile tea; prepare an infusion using 2 to 3 grams powdered chamomile in 8 ounces of boiling water or add 3 to 5 milliliters tincture to hot water. Take either preparation three times daily between meals.

AN OUNCE OF PREVENTION

Dietary changes can help to prevent some outbreaks of irritable bowel syndrome and inflammatory bowel disease. It is important to identify and avoid potential food allergens or irritants. Studies have found that approximately two-thirds of people with IBS have at least one food allergy; the most common culprits are dairy products and grains.

CALL FOR HELP

If you have irritable bowel syndrome or inflammatory bowel disease, you should visit your doctor for an accurate diagnosis. Symptoms of bowel disorders could possibly indicate other, more serious conditions.

Macular Degeneration

Macular degeneration is a serious eye disorder in which the macula, a tiny portion of the retina in the back of the eye, is damaged, causing blindness in the central vision while leaving peripheral vision intact. The damage to the retina can be caused by or associated with various factors, including smoking, sunlight, diabetes, high blood pressure, atherosclerosis, and heart disease.

Approximately 13 million Americans have macular degeneration in some stage of development. This visual disorder typically affects people over age fifty-five, and the risk of getting this condition increases with age. Danger to the eye most often takes the form of deposits that build up under the macula or, less often, of an abnormal growth of blood vessels that leak fluid into the retina. Macular degeneration cannot be reversed, but it can be halted and risk for the disorder can be greatly reduced. It's been shown, for example, that people who have high blood levels of antioxidants have a lower risk of developing macular degeneration.

MOST HELPFUL SUPPLEMENTS

- *Antioxidants*—vitamins A (beta-carotene), C, E, and selenium—help fight free-radical damage in the eyes. Take 10,000 IU beta-carotene, 1,500 milligrams vitamin C in divided doses, 400 IU of vitamin E, and 200 micrograms of selenium daily.

- *Bilberry* strengthens and reinforces the collagen in the retina. Take standardized extract or capsules daily, following package directions.

- *Ginkgo* reduces the risk of developing macular degeneration. Take 120 milligrams standardized extract daily or ½ teaspoon of tincture three times daily.

AN OUNCE OF PREVENTION

The steps you can take to prevent macular degeneration are the same used to prevent atherosclerosis (see page 187). These steps will protect against free-radical damage and improve the supply of oxygen and blood to the macula.

CALL FOR HELP

If you detect any change in your visual acuity, contact your doctor or ophthalmologist. After age fifty, it is important to have annual eye exams to detect macular degeneration—and other eye diseases—as early as possible to avoid permanent vision damage.

Menopausal Symptoms

Some women glide through menopause with few complaints, while others must endure a number of difficult physical and psychological adjustments as they work their way through the "changes of life."

At the most basic level, menopause is nothing more than the cessation of ovulation and menstrual cycles. It usually occurs in women between the ages of forty-five and fifty-five, with the average age being fifty-one. Most women experience irregular periods for five to seven years before their cycle stops entirely. (This erratic time is known as perimenopause.) A women is considered to have passed through menopause after going one full year without menstrual periods.

Menopausal symptoms—including hot flashes, vaginal dryness, and anxiety—occur in response to changes in estrogen and progesterone levels in the body. Hot flashes occur when the estrogen levels drop, causing a sudden adjustment in the body's thermostat and an abrupt "flash" of heat. (The pituitary gland in the brain controls both estrogen levels and body temperature, hence the link between hormones and heat.)

Typically, the heat of a hot flash begins in the chest and spreads to the neck, face, and arms. It can be accompanied by sweating and heart palpitations, and it may be followed by chills. Three out of four menopausal women experience hot flashes, which can occur as often as once an hour and can last for three or four minutes at a stretch. For most women, hot flashes are mild and end within two years, but 25 percent of women who experience hot flashes suffer from them for more than five years, and about 10 percent "flash" for the rest of their lives.

In addition to the physical changes, many women experience mood swings, depression, insomnia, and irritability during menopause. These psychological disturbances can be caused by organic changes and shifting hormone levels, but they can also be exacerbated by the other life changes taking place during the late forties and early fifties, such as children leaving home and career-related stresses.

The discomforts of menopause can be especially severe if menstruation stops abruptly, either naturally or following the surgical removal of the ovaries. Whatever the trigger event, natural remedies can be very useful in treating symptoms of menopause.

MOST HELPFUL SUPPLEMENTS

- *Black cohosh* contains key ingredients (triterpenes and flavone derivatives) that act on the hypothalamus and vasomotor centers in the brain. It has been shown to have better results than estrogen at relieving menopausal complaints. In one double-blind study, 80 women were given either black cohosh extract, estrogen (Premarin), or a sugar pill for 12 weeks. The black cohosh was the most effective at reducing menopausal complaints, lowering number of hot flashes from an average of five to less than one, compared with lowering the number to 3.5 for the estrogen group. Commercial products are available; follow package directions.

- *Chaste berry* helps balance the hormonal system. Take 500 to 1,000 milligrams daily. Commercially prepared tinctures are available; follow package directions.

- *Vitamin E* helps eliminate hot flashes. Take up to 400 IU daily.

- *Wild yam* roots contain diosgenin, which can be converted in a laboratory to estrogen and progesteronelike compounds. In fact, until 1970, this plant was used in the manufacture of birth control pills. While the body does not convert wild yam into estrogen, it does use the active ingredients to relieve menopausal symptoms. This herb is commercially available; follow package directions.

- *Soy* and other foods rich in soy protein contain phytoestrogens, or molecules very similar to estrogen and progesterone. A menopausal woman who eats lots of soy protein can retain the benefits she had previously received from her body's natural estrogen. Try to consume 25 grams of soy daily; a soy burger contains about 18 grams, a glass of soy milk about 8 grams. Powdered soy supplements are also available; follow package directions.

OTHER HELPFUL SUPPLEMENTS

- *Calcium* helps relieve nervousness and irritability. In addition, calcium helps ward off osteoporosis, or thinning of the bones, which begins at menopause (see page 287). Take 1,500 milligrams daily in divided doses.

- *Evening primrose oil* relieves hot flashes and helps in the production of estrogen. Take 500 milligrams twice daily.

- *Fenugreek* contains diosgenin, a chemical similar to the female hormone estrogen. The herb is often used in the treatment of hot flashes and depression associated with menopause. Fenugreek tea and commercial products are available; follow package directions.

- *Dandelion* has mild diuretic effects, which can help to reduce menopausal water retention. Dandelions can be eaten fresh in salads. Dandelion tea and capsules are available; follow package directions.

- *Dong quai* is useful in treating menopausal symptoms. Commercial products are available; follow package directions.

AN OUNCE OF PREVENTION

Menopause is a natural—and inevitable—part of the aging process. It is an important stage of a woman's reproductive cycle and need not be feared or met with apprehension.

CALL FOR HELP

One out of every two women experiences some symptoms associated with menopause, and about one in four finds those symptoms uncomfortable to distressing. If the symptoms of menopause interfere with your quality of life or if you would like to discuss the pros and cons of hormone replacement therapy, consult your doctor.

Nausea and Morning Sickness

Nausea, or feeling "sick to your stomach," is a common condition that can be caused by a variety of circumstances and conditions. Feelings of nausea can be caused by food allergy, gallstones, diabetes, constipation, or because you have a viral infection. Nausea associated with motion sickness is often seen in both children and adults. Another form of nausea is morning sickness, which often occurs in women during their first three months of pregnancy.

The suggested supplements listed below can relieve nausea; however, if you have frequent bouts of nausea, the cause also needs to be addressed to prevent recurrence and to make sure you do not have another underlying health problem.

MOST HELPFUL SUPPLEMENTS

- *Ginger* relieves symptoms of nausea, motion sickness, and morning sickness. In addition, studies have found ginger to be quite effective in treating nausea associated with chemotherapy. Take one of the following three times daily: 250-milligram capsules or tablets; 1 cup ginger infusion (250 milligrams ginger powder in 8 ounces of boiling water); or ½ teaspoon of tincture. Take on an empty stomach.

- *Peppermint* relieves nausea. Drink up to 3 cups of tea daily.

- *Vitamin B6* (pyridoxine) helps relieve nausea. For morning sickness, take 10 to 25 milligrams three times daily. For motion sickness, take 100 milligrams one hour before the trip, followed by 100 milligrams two hours later.

- *Acidophilus* helps restore bacterial balance in the digestive tract. Take one dose three times daily for a week; follow package label for dosage information.

OTHER HELPFUL SUPPLEMENTS

- *Licorice* root tea helps soothe an unsettled stomach. Take one dose up to three times daily; follow package directions.

AN OUNCE OF PREVENTION

To decrease the likelihood of picking up a virus or bacteria that can cause nausea, practice good hand-washing habits. Also try to avoid overeating and eating too quickly, which can cause stomach upset. Identify any possible food allergies, which may contribute to feelings of nausea after eating.

CALL FOR HELP

The occasional bout of nausea is unpleasant, but not harmful. If you experience chronic nausea and you are not pregnant, contact your doctor if the symptoms persist for several weeks. As noted earlier, persistent nausea can be a symptom of more serious medical problems.

If you begin to vomit, look for signs of dehydration, including dark yellow urine, dry mucous membranes, drowsiness, and skin that doesn't bounce back right away when you pinch it. If you experience vomiting along with fever, bloating, if the vomit looks like coffee grounds, or if vomiting follows a head injury, seek prompt medical attention.

Obesity

It's not easy being fat: Obesity takes both an emotional and a physical toll on all those who suffer from it. If you are more than 20 percent above the normal weight for your age, height, and body frame, you are considered obese. Another way to assess obesity is to measure the percentage of body fat. Women whose body fat is greater than 30 percent and men with a percentage greater than 25 percent are considered to be obese.

Those extra pounds also put you at greater risk of a number of health problems, including heart disease, diabetes, high blood pressure, stroke, gallbladder disease, hemorrhoids, varicose veins, gallstones, certain cancers, and kidney and liver problems. Excessive weight also contributes to infertility, osteoporosis, varicose veins, and PMS symptoms. Many natural substances can help people lose weight; however, a healthy diet, regular exercise, and a positive attitude are the mainstays of a successful weight-loss program.

Approximately one out of every three American adults and one out of five American children is obese. Although obesity was once thought to be caused simply by overeating, researchers now know that it is a more complex issue. Several factors may contribute to any one person's being overweight, including lack of adequate physical activity, consuming a nutritionally poor diet, low serotonin levels in the brain, impaired metabolism, glandular malfunctions, sensitivity to insulin, and heredity.

MOST HELPFUL SUPPLEMENTS

- *Chromium* increases fat metabolism. In one study, volunteers at a Texas weight loss center were given either chromium supplements or a placebo for seventy-two days; they were not given any diet or exercise regimen. Those taking chromium lost an average of 4.2 pounds of fat and gained 1.4 pounds of lean muscle mass. There was little change in those taking sugar pills. Other studies have found the benefits of chromium supplementation to be greatest in people with chromium deficiency as well as the elderly. Take 200 to 400 micrograms of chromium picolinate daily.

- *Coenzyme Q10* promotes weight loss. Take 100 to 300 milligrams daily.

- *5-HTP* increases serotonin levels, which reduces hunger and promotes weight loss. Take 100 milligrams twenty minutes before each meal for two weeks.

OTHER HELPFUL SUPPLEMENTS

- *Carnitine* helps metabolize fat and reduce feelings of hunger. Take 1 to 3 grams daily.

- *DHEA* increases fat metabolism. Take 30 to 90 milligrams daily.

- *Psyllium* reduces fat absorption. Stir ½ to 1 teaspoon psyllium powder or husks in 8 ounces of water and drink 2 cups per day. Do not take any other supplements or medications at the same time.

AN OUNCE OF PREVENTION

You can avoid excessive weight gain by shedding extra pounds before things get out of hand. When the bathroom scale shows you weigh five pounds more than your goal weight, it's time to watch your diet and exercise habits.

Also, be sure to drink at least eight 8-ounce glasses of water a day. The water helps keep the body hydrated, it helps with metabolism—and it's filling. Try drinking a glass of water before you sit down to a meal if you feel you might be tempted to overindulge. Eat slowly; it can take twenty minutes for your brain to recognize that your stomach feels full.

CALL FOR HELP

It can be exceedingly difficult to lose weight. Don't feel you have to battle the bulge alone. Make an appointment with your doctor or a dietician for help working out a weight-loss plan.

Osteoporosis

No matter what your age, it's time to begin battling osteoporosis. Osteoporosis, a medical condition that literally means "porous bones," affects about 25 million Americans—at least one out of every three women over age sixty. (It is much less common and less severe in men, in part because their bones are one-third denser than women's to begin with.) The thin, brittle bones associated with osteoporosis can lead to broken hips, vertebrae, and other bones. In advanced cases, a strong cough can break a rib, or a gentle bump can cause a fractured hip. In fact, nearly half a million older people take a fall each year that results in a broken hip caused at least partly by osteoporosis.

In most cases, the problem arises when people have consumed too little calcium. The body needs this essential mineral for muscle contractions and other functions, so it "steals" from the bones, leaving them fragile and thin. Typically, people experience their peak bone mass in the spine at around age thirty and in the long bones at around age thirty-five. After that, bone mass drops by about 1 percent a year, until menopause.

After menopause, bone loss speeds up to about 2 to 4 percent a year for the next ten years or so. (For every 10 percent loss in bone mass, your risk of bone fracture doubles.) Bone loss speeds up at menopause because the body has less estrogen, which helps the body absorb and use available calcium, and because the body produces less calcitonin, the hormone that prompts the bones to absorb calcium. By the age of eighty, most women have lost between a quarter and half their bone mass.

As the disease progresses, the spinal column may become compressed, causing the appearance of "shrinking." The spine may also curve, resulting in the characteristic "dowager's hump." These spinal changes actually result from fractures caused by the pressure of the body's weight on weak and brittle bones.

In addition to dietary calcium deficiency, osteoporosis can also be caused by the inability to absorb enough calcium through the intestine, a calcium-phosphorus imbalance, a lack of exercise, and prolonged jaundice. Women at risk include those with a thin frame, sedentary lifestyle, and a family history of the disease. Smokers, alcoholics, diabetics, women who reach menopause before age forty, those who consume a lot of caffeine, women who have never been pregnant, Asian women, and white women (especially blondes and redheads of Northern European ancestry) are also at increased risk. Certain drugs, such as cortisone, anticoagulants, anticonvulsants, and thyroid medications, can also contribute to calcium loss.

Osteoporosis is very often preventable if a healthy diet and lifestyle habits are followed. The treatment options listed here help prevent osteoporosis and help slow its progression.

MOST HELPFUL SUPPLEMENTS

* *Boron* improves absorption of calcium. A study conducted by the U.S. Department of Agriculture indicated that within eight days of supplementing their daily diet with 3 milligrams of boron, a test group of postmenopausal women lost 40 percent less calcium, one-third less magnesium, and slightly less phosphorus through their urine than they had before beginning boron supplementation. Take 3 to 5 mil-

ligrams daily as sodium tetrahydroborate or sodium tetraborate decahydrate.

- *Calcium* is necessary for bone formation. Building bone density through calcium supplementation before menopause can help delay osteoporosis later in life. In a two-year study, 214 perimenopausal women received either 1,000 or 2,000 milligrams of calcium. While the control group actually lost 3.2 percent of their spinal bone density, the calcium-treated groups increased their density by 1.6 percent (there was no difference between the two calcium groups). In postmenopausal women, long-term studies have demonstrated that calcium supplementation does slow the rate of calcium loss down by at least 30 to 50 percent and offers significantly greater protection against hip fractures. Take 1,500 to 2,000 milligrams daily, along with 750 to 1,000 milligrams of magnesium. *Hint*: You can also boost the calcium content of many dishes without increasing fat by adding powdered nonfat dry milk to recipes; every teaspoon of powdered milk adds about 50 milligrams of calcium.

OTHER HELPFUL SUPPLEMENTS

- *Phosphorus* improves bone formation. Take 99 milligrams daily.

- *Vitamin D* is necessary for calcium uptake. People who do not get sufficient exposure to the sun may benefit from vitamin D supplementation. A good multivitamin–mineral supplement should contain 400 IU daily, which is all you need.

- *Zinc* improves calcium uptake. Take 50 milligrams daily.

AN OUNCE OF PREVENTION

The best way to prevent osteoporosis is to build strong bones early in life and then take steps to keep bones healthy later in life. To minimize bone thinning, add calcium to your diet, either by eating more calcium-rich foods or by taking a supplement. The average American diet contains just 500 milligrams of calcium, but after menopause a woman needs about 1,500 milligrams.

In addition, build your bones through regular weight-bearing exercise, such as walking, bicycle riding, or dancing. Don't smoke or drink alcohol in excess, both of which can contribute to brittle bones.

CALL FOR HELP

Osteoporosis is a silent disease; it usually doesn't provide much advance warning. In fact, the first sign is often a severe hip or low-back pain, or a broken bone after a minor bump or fall. If you have sharp unexplained back pain that doesn't improve after two or three days, see your doctor.

Parkinson's Disease

Parkinson's is a paradoxical disease. When people suffer from it, some of their muscles become rigid and others contract involuntarily. People with Parkinson's disease may stoop, shuffle, and present a vacant, mask-like, expressionless face, while at the same time suffer from an incessant tremor in the hand.

Parkinson's disease involves a failure of the body's internal communication system. When we are healthy, we take our bodies for granted, but every move we make, from kicking a ball or writing our names, requires thousand of coordinated communications between the brain, muscles, tendons, and bones. When these systems work well together, we think nothing of it, but when some part of the network breaks down, the effect can be devastating, as it is with Parkinson's disease.

In 1817 British physician James Parkinson identified the disease that bears his name, though it was first called simply the "shaking palsy." The disease, which afflicts more than 1 million Americans, involves damage to the middle section of the brain known as the *substantia nigra,* named for its blackish pigmentation. This midbrain area is the main supplier of dopamine, the neurotransmitter that allows for communication about movement between various parts of the body. When these cells die off and the dopamine supply dwindles, the nerve signals cross and muscle action goes haywire.

Parkinson's is a degenerative disease that usually first shows up when patients are in their fifties and sixties. There is no known cure for the disease, but the symptoms can be relieved through medication and natural treatments. Early treatment can help to slow the progress of the disease.

The cause of Parkinson's disease is unknown, though some experts suspect that a virus, malnutrition, or chemical exposure could be involved. Supporting evidence for the virus-trigger theory is that many people who survived the encephalitis epidemics between 1919 and 1926 (caused by a virus) developed Parkinson's years later. The toxin-trigger theory was bolstered by evidence of an outbreak of Parkinson's-like disorders among drug addicts in San Francisco in the early 1980s.

In some cases people develop symptoms of Parkinson's disease that prove actually to be side effects of medication. This is called Parkinson's syndrome rather than Parkinson's disease, and the symptoms disappear when the drugs are discontinued. If you suspect you have Parkinson's syndrome, review your use of all prescription and nonprescription drugs, and discuss the issue with your doctor.

MOST HELPFUL SUPPLEMENTS

- *Calcium* and *magnesium* are needed for nerve signal transmission. Take 1,500 milligrams, along with 750 milligrams magnesium.

- *Lecithin* is needed for nerve signal transmission. Commercial products are available; follow package directions.

- *Vitamin B6* (pyridoxine) is necessary for the production of dopamine. Take up to 1,000 milligrams daily.

- *Vitamin B complex* is critical for brain function. Take 100 milligrams three times daily with meals.

OTHER HELPFUL SUPPLEMENTS

- *Vitamin C* can slow progression of disease in people not yet taking medication; it also improves cerebral circulation. Vitamin C is essential for the manufacture of certain nerve-impulse-transmitting substances and hormones. A study showed that 60 percent of the elderly people with vitamin C deficiency had Parkinson's disease. Take 3,000 milligrams daily, in divided doses.

- *Vitamin E* slows progression of disease. Take 400 IU daily.

AN OUNCE OF PREVENTION

It's impossible to prevent Parkinson's because we do not understand its causes. Some experts suspect that environmental toxins and pesticides may play a role, but this link is not well established.

CALL FOR HELP

Unfortunately, the early warning signs of Parkinson's include vague, minor symptoms that are often written off as typical signs of aging, including fatigue, stiffness, difficulty swallowing, and slight hand tremor. Classic Parkinson's symptoms—such as a constant "pill-rolling" motion of the fingers, a tendency to hold an arm with the elbow bent, small handwriting, and a "mask-like" expression—tend to show up next. Finally, more severe signs appear, such as a slow, shuffling walk, severe tremor, stooped posture, muscle rigidity, and dementia. If you experience any symptoms of Parkinson's—however mild—be sure to discuss the issue with your physician at your annual physical.

PMS (Premenstrual Syndrome)

An estimated 40 percent of women experience PMS—premenstrual syndrome—to some degree each month. PMS is not a disease, but a combination of symptoms that occurs seven to fourteen days before menstruation begins.

Symptoms may include fatigue, irritability, depression, headache, nervousness, anxiety, mood swings, abdominal bloating, diarrhea, constipation, cravings for sugar, tender and enlarged beasts, uterine cramping, altered sex drive, backache, acne, and swelling of the ankles and fingers. These symptoms are the result of fluctuations in a woman's hormone levels (estrogen and progesterone) prior to menstruation.

MOST HELPFUL SUPPLEMENTS

The long list of effective options for PMS reflects the many symptoms associated with the syndrome. Choose the options that match your symptoms.

- *Black cohosh* reduces cramping, depression, and mood swings. Commercial products are available; follow package directions.

- *Calcium* has been found to improve many PMS symptoms characteristic of PMS. A double-blind study of 500 women with PMS found that supplementation with 1,200 milligrams of calcium (as calcium carbonate) over a three-month period was very effective in reducing moodiness, water retention, food cravings, and pain; fully 73 percent of women in the study reported fewer PMS symptoms. Take 1,500 milligrams daily in divided doses.

- *Magnesium* helps relieve pain and nervousness. Studies have found that 45 percent of women with PMS have low levels of magnesium. Other studies have found that women taking magnesium supplements reported significant reductions in "menstrual distress." Take 12 milligrams per 2.2 pounds of body weight (e.g., a 110-pound woman would take 600 milligrams).

- *Vitamin B6* (pyridoxine) relieves depression and helps magnesium enter the cells. Take 50 milligrams one to two times daily.

- *Vitamin E* relieves breast tenderness, headache, fatigue, nervous tension. In a study of 75 women treated for two months with vitamin E, 75 percent reported improvement in their PMS symptoms. Take 400 IU daily.

OTHER HELPFUL SUPPLEMENTS

- *Chaste berry* helps balance hormone levels. Commercial products are available; follow package directions.

- *Dandelion* is a mild diuretic. Take ½ teaspoon of tincture three times daily to minimize bloating.

- *Evening primrose oil* relieves cramps, depression, breast tenderness, and other PMS symptoms. Take 500 milligrams twice daily.

- *Licorice* lowers estrogen and raises progesterone levels; it also reduces water retention. Take any of the following forms three times daily: 1 to 2 grams powdered root as tea, 4 milliliters fluid extract, or 250 to 500 milligrams of solid dry root.

- *Dong quai* relieves hot flashes and cramps. Begin taking the herbs on day fourteen of your menstrual cycle until menstruation begins. Commercial products are available; follow label directions for dosage information.

AN OUNCE OF PREVENTION

You may be able to ease the symptoms of PMS by consuming a diet high in fiber and complex carbohydrates, eating small meals with snacks in between. This will help stabilize blood sugar, which can minimize

PMS symptoms. Eat salads and lighter foods; avoid spicy and acidic foods. Avoid caffeine.

CALL FOR HELP

If your PMS symptoms interfere with your relationships and quality of life, talk to your doctor about it. You may need some clinical tests to rule out other hormone imbalances that may be contributing to your condition.

Prostate Enlargement (BPH)

As men age, many develop problems with their prostate, a walnut-sized gland located just below the bladder and which wraps around the urethra (the tube through which semen and urine leave the body). In some cases, the prostate becomes enlarged due to stimulation by the hormone dihydrotestosterone. This enlargement can disrupt the flow of urine when the prostate presses against the urethral opening. When this occurs, a man is said to have a noncancerous condition known as benign prostate hyperplasia (BPH); it affects about half of all men over age 50.

Untreated, the urethra may eventually become blocked and cause kidney damage. Symptoms of BPH include increased urinary frequency, urgency, a need to get up at night to urinate, reduced urinary flow, and difficulty emptying the bladder.

Another prostate condition that often affects men is prostatitis, which is an infection or inflammation of the prostate. Symptoms include pain during urination, discharge from the penis, and fever. Prostatitis can affect men of any age. It should be treated by a doctor.

MOST HELPFUL SUPPLEMENTS

- *Saw palmetto* blocks the formation of dihydrotestosterone, improving urinary flow. Take 160 milligrams twice daily of extract standardized to 85 to 95 percent fatty acids and sterols. Commercial products are available; follow package directions.

- *Zinc* deficiency has been linked to prostatitis. Take 45 to 60 milligrams daily.

OTHER HELPFUL SUPPLEMENTS

- *Nettle* often improves symptoms of BPH. Take 300 to 600 milligrams of the root extract daily.

- *Vitamin B6* (pyridoxine) aids absorption of zinc. Take 50 milligrams daily.

AN OUNCE OF PREVENTION

BPH is a common side effect of aging. The nutritional supplements listed above may help to prevent BPH, but there is little else you can do to avoid it.

CALL FOR HELP

If you experience the symptoms of prostatitis—pain during urination, discharge from the penis, and fever—contact your doctor. You may need antibiotics or other medicines to clear the condition. In addition, if you experience urinary difficulty that affects your quality of life, discuss the matter with your doctor; in some cases, surgery is required to treat BPH.

Psoriasis and Seborrheic Dermatitis

Psoriasis is a noncontagious skin disorder caused by a defect in the skin cells that allows them to reproduce much more rapidly than normal. It is characterized by patches of red or salmon-colored skin, often with thick, silvery scales on top. The condition can occur on any part of the body, but it is most common on the elbows and knees. Itching is present in about 30 percent of cases.

On a biochemical level, psoriasis is caused by two substances responsible for controlling this cell activity—cyclic adenosine monophosphate (AMP) and cyclic guanidine monophosphate (GMP). When a person has high levels of GMP and low levels of AMP the result is psoriasis. Approximately four to six percent of all Americans suffer with psoriasis; it affects men and women equally, but is much more common among whites than blacks or Native Americans. It appears to have a genetic link, although the exact cause of psoriasis is unknown.

Seborrheic dermatitis (seborrhea) is a common skin disorder that is characterized by oily patches of skin that form crusts and scales. This skin problem is caused by a malfunction of the sebaceous glands, which secrete oil. Seborrheic dermatitis has several forms, including nigra (dark-colored patches), sicca (dry with scales), facier (appears on the face), and rosacea (condition that first appears in middle age and often reappears).

The cause of seborrheic dermatitis is unknown. Possible causes include: vitamin deficiency (vitamin A and/or vitamin B; biotin deficiency in infants), hormone irregularities, food deficiency of vitamin B6 (pyridoxine), taking certain drugs (e.g., oral contraceptives, dopamine), and exposure to FD&C yellow #5 dye.

MOST HELPFUL SUPPLEMENTS

- *Milk thistle* improves liver function and eases the symptoms of psoriasis. Take 100 milligrams of standardized extract three times daily.

- *Vitamin A* is vital for healthy skin. Take 10,000 IU as beta-carotene daily.

- *Vitamin B complex* is an antistress vitamin that also helps maintain healthy skin. Take 50 milligrams three times daily.

- *Omega-3 fatty acids* nourish the skin. Take 1 tablespoon oil daily.

- *Evening primrose oil* promotes healing and helps prevent dryness. Take 2 grams daily in divided doses.

OTHER HELPFUL SUPPLEMENTS

- *Chromium* increases sensitivity of receptors to insulin, a hormone that is typically high in people with psoriasis. Take 400 micrograms daily.

- *Oatmeal* baths can minimize itching associated with psoriasis. Use a commercial product; follow package directions.

- *Flaxseed* can help restore normal oil production. Take 1 tablespoon of oil daily. Commercial products are available; follow package directions.

- *Vitamin C* is essential for healthy collagen and connective tissue. Take 1,500 milligrams daily in divided doses.

AN OUNCE OF PREVENTION

Poor digestion or absorption of proteins and a diet low in fiber are typical of people with psoriasis. Eat a high-fiber diet and increase your intake of omega-3 fatty acids (found in salmon, mackerel, and other cold-water fish). Avoid alcohol, which can significantly worsen psoriasis.

CALL FOR HELP

With psoriasis or seborrheic dermatitis, it is best to visit a dermatologist for an initial diagnosis and treatment plan. Supplements and natural remedies may prove useful for subsequent outbreaks and for prevention. In addition, contact your doctor if you experience any signs of skin infection, such as swelling, redness, discharge (pus), and fever.

Ulcers

At first you may have thought it was a bad case of heartburn or indigestion. But when the burning pain failed to subside, you suspected much worse: an ulcer.

At least five million Americans suffer from ulcers, or open sores or lesions in the lining of the gastrointestinal tract. Ulcers form when there is too much stomach acid or when the mucous lining fails to protect the digestive tract. People who take large doses of stomach-irritating antiinflammatory drugs to manage pain are at increased risk of developing ulcers.

There are two kinds of ulcers: gastric, which appear in the stomach, and duodenal, which form in the duodenum (the first part of the small intestine). Duodenal ulcers are about five times more common than gastric ulcers and about four times more common in men than in women.

Researchers have found that many ulcers are caused or exacerbated by a bacterium known as *Helicobacter pylori*. A two-week regimen of antibiotics can wipe out this bacterium, vastly diminishing the odds of a recurrence. Research shows that the bacterium is present in up to 100 percent of patients with duodenal ulcer and in up to 70 percent of those with gastric ulcer. Factors that predisposed to infection by *H. pylori* include low levels of antioxidants in the stomach and intestinal linings and low gastric acid production.

Left untreated or undetected, ulcers can lead to peritonitis, an inflammation of the lining of the abdominal cavity. In severe cases, an ulcer can burn a hole right through the stomach or intestine; this is a medical emergency. Fortunately, a number of natural remedies can be used to prevent and treat ulcers before they reach this stage.

Symptoms of ulcer include abdominal discomfort (usually described as burning, gnawing, or cramplike) forty-five to sixty minutes after eating or during the night. This distress is relieved by food, antacids, or vomiting. The abdomen may be tender, and some patients have blood in their stools.

MOST HELPFUL SUPPLEMENTS

- *Licorice* or DGL (deglycyrrhizinated licorice), an extract of the licorice herb, helps heal ulcers. A three-month study of duodenal ulcer sufferers found that DGL healed ulcers faster than the popular drug Tagamet. Commercially prepared DGL products are available; follow package directions.

- *Psyllium* provides supplemental fiber. Take as needed. Begin with ½ teaspoon psyllium seeds or powder mixed in 8 ounces of cool liquid and drink 2 to 3 cups daily. Stir the mixture vigorously and drink it quickly, followed by additional water. Do not take psyllium if you are taking supplemental fiber.

- *Zinc* promotes healing of the digestive tract. Take 20 to 30 milligrams daily.

- *Fiber* is associated with a reduced rate of duodenal ulcer; it also cuts recurrence rates of the illness. Fiber promotes the secretion of mucus, which helps protect the digestive tract. Commercial products are available; follow package directions.

OTHER HELPFUL SUPPLEMENTS

- *Aloe vera* relieves pain and promotes healing. Drink 30 milliliters aloe gel or juice three times daily or take 50 to 200 milligrams of powder or powdered capsules.

- *Vitamin C* helps rebuild antioxidant levels. Take 500 milligrams three times daily.

- *Vitamin E* helps rebuild antioxidant levels. Take 100 IU three times daily.

AN OUNCE OF PREVENTION

To minimize the risk of ulcers, eat six small meals a day rather than three larger ones. The presence of food can help to neutralize stomach acid, so don't go too long without eating. Eliminate alcohol, and minimize your intake of caffeine. Also take steps to manage stress, which can increase production of stomach acid. Quit smoking; tobacco smoke constricts the blood vessels lining the stomach, making the stomach wall more vulnerable to sores. Avoid the use of aspirin and other nonsteroidal antiinflammatory drugs, which are associated with the formation of ulcers.

CALL FOR HELP

Seek medical care immediately if you vomit blood or material that looks like coffee grounds. Also seek help if black, tarry blood appears in the stool. These are signs that the ulcer is bleeding, which can cause anemia. Blood can also indicate a perforated ulcer, one that has burned through the stomach or intestinal wall.

Urinary-Tract Infections

There are three common types of urinary-tract infections: *urethritis* (infection of the tube that leads from the bladder to the outside of the body), *cystitis* (infection of the bladder), and *pyelitis* (infection of the kidneys). The urinary system is interconnected, so infection tends to spread. The condition should be treated at the first sign of problems.

Urinary-tract infections are typically caused by bacteria that enter the bladder through the urethra, then travel to the bladder and kidneys. Cystitis occurs much more often in women than in men, primarily because women have a shorter urethra than do men, and the anus and vagina openings are close to the urethra opening, which allows bacteria easy access to the urethra. The bacteria that most often cause cystitis and kidney infections are *E. coli*, which reside in the intestinal tract. Two sexually transmitted bacteria, *mycoplasma* and *chlamydia*, also cause cystitis. Urinary-tract infections can also result from use of a catheter (a tube inserted into the bladder to empty it).

Symptoms of urinary-tract infections include urgent

and frequent need to urinate. Urination is often painful and described as a burning sensation, and the urgency may persist even after emptying the bladder. The urine is often cloudy, with an unpleasant odor.

MOST HELPFUL SUPPLEMENTS

- *Acidophilus* restores the population of "good" bacteria in the intestinal tract. Take 2 capsules three times daily.

- *Goldenseal* prevents bacteria from adhering to the intestinal walls. Take any one of the following three times daily: 4 to 6 milliliters (1 to 1½ teaspoons) tincture; 250 to 500 milligrams of standardized root in capsules or tablets, or 0.5 to 2.0 milliliters (¼ to ½ teaspoon) fluid extract.

- *Uva ursi* is antiinflammatory, antibacterial, and a diuretic agent. Commercial products are available; follow product directions.

- *Vitamin C* acidifies the urine, which helps eliminate the bacteria. Take 1,500 or 3,000 milligrams daily in divided doses. *Note*: If you are also taking an antibiotic, check with your physician before taking vitamin C, as it may interfere with the effectiveness of the antibiotic.

OTHER HELPFUL SUPPLEMENTS

- *Nettle* is an antiinflammatory. Steep 1 teaspoon dried, crushed nettle leaves or root in 8 ounces of boiling water. Allow the mixture to cool and drink 1 tablespoon every hour up to 8 ounces per day.

- *Garlic* is a natural antibiotic. Take one to two doses of odorless garlic daily until symptoms subside.

- *Zinc* boosts the immune system. Take 30 milligrams daily.

AN OUNCE OF PREVENTION

To prevent urinary-tract infections, always wipe from front to back after bowel movements to avoid fecal contamination of the urethra. Girls should use white, unscented toilet paper and mild soaps when bathing.

CALL FOR HELP

If painful urination is accompanied by fever, chills, bloody urine, vomiting, back or abdominal pain, get medical help immediately, as these are symptoms of kidney disease.

Varicose Veins

Varicose veins are nothing to be ashamed of, even though many people go to great lengths to hide them. In fact, more than 40 million Americans live with the swollen and sometimes painful veins, which most often appear in the legs.

Gravity is to blame for the formation of varicose veins. Blood that circulates to the legs must be pumped uphill to the heart, against the pull of gravity. Veins are equipped with one-way valves to prevent the blood from flowing back down the legs, but when the valves are stretched or damaged, they don't close properly, and the blood slips back down and pools, causing the veins to stretch out and appear blue and puffy.

Women experience varicose veins about four times more often than men, partially because of the rigors of pregnancy and childbirth. In preparation for childbirth a pregnant woman's body releases hormones that weaken the collagen and connective tissues in the pelvis to make the birth process easier. But these hormones can also weaken the collagen and valves in the veins, increasing the likelihood of varicose veins. Other common causes of varicose veins include standing for long periods (the pressure exerted against the veins can increase up to ten times when standing), and sitting for long periods without movement, especially with the legs crossed. Genetics and obesity also contribute to the problem.

While some people consider varicose veins unsightly, they do not pose a health risk in most cases, when the affected veins are near the skin surface. Varicose veins that form deep within the leg, however, can lead to more serious complications, such as skin ulcers, phlebitis (inflammation of the vein), and thrombosis (formation of a blood clot). These conditions require immediate medical attention.

MOST HELPFUL SUPPLEMENTS

• *Bromelain* promotes breakdown of fibrin, the protein that forms during the blood's clotting response. Take 500 to 750 milligrams two to three times daily between meals.

• *Vitamin C* promotes healthy veins. It also aids circulation by reducing the tendency for blood clots to form. Take 1,500 milligrams daily in divided doses.

• *Zinc* promotes healthy veins. Take 15 to 30 milligrams daily.

OTHER HELPFUL SUPPLEMENTS

- *Bilberry* strengthens vein walls. Use the tea externally as a wash on the affected area. You may also use a commercial product; follow package directions.

- *Fiber* can be used to build bulk in the stool and reduce straining during bowel movements. Over time, the straining of constipation can weaken the vein walls, resulting in varicose veins or hemorrhoids (varicose veins in the anus). Use a commercially prepared product; follow package directions.

AN OUNCE OF PREVENTION

You can reduce your risk of developing varicose veins by avoiding standing for long periods of time, crossing your legs, lifting heavy objects, and wearing tight shoes, garters, or undergarments. Regular exercise (especially walking, biking, and jogging) helps to prevent varicose veins and to combat them once they have formed. The contraction of the leg muscles helps move the pooled blood in the legs back into circulation.

CALL FOR HELP

If you experience sharp leg pans or notice a red lump in the vein that doesn't go away when you put your legs up, contact your doctor immediately. You may have a blood clot, which can lead to serious medical problems, including stroke.

∿ ORGANIZATIONS OF INTEREST

NATUROPATHY

A licensed naturopathic physician (N.D.) is a graduate of a four-year graduate-level naturopathic medical school. Naturopaths learn the same basic sciences that traditional medical doctors do, but their training includes additional coursework in herbal medicine, nutrition, homeopathy, and exercise therapy. Currently, the District of Columbia and thirteen states—Alaska, Arizona, California, Connecticut, Hawaii, Kansas, Maine, Montana, New Hampshire, Oregon, Utah, Vermont, and Washington—require that naturopathic physicians pass a state licensing exam.

For a referral to a naturopathic physician in your area, contact:

THE AMERICAN ASSOCIATION OF NATUROPATHIC
PHYSICIANS
3201 New Mexico Avenue, NW
Suite 350
Washington, DC 20016
Toll-free (866) 538-2267; (202) 895-1392
www.naturopathic.org

HOLISTIC MEDICINE

Holistic medicine is practiced by medical doctors (M.D.s), osteopaths (D.O.s), and naturopaths (N.D.s). These physicians emphasize the treatment of the whole person and encourage personal responsibility for health.

For a referral to a licensed holistic practitioner, contact:

THE AMERICAN HOLISTIC MEDICAL ASSOCIATION
12101 Menaul Blvd. NE
Suite C
Albuquerque, NM 87112
(505) 292-7788
www.holisticmedicine.org

THE AMERICAN HOLISTIC HEALTH ASSOCIATION
P.O. Box 17400
Anaheim, CA 92817-7400
(714) 779-6152
www.ahha.org

HERBAL MEDICINE

Herbal medicine is used by many naturopathic physicians and a growing number of conventional doctors. There is no separate certification or licensing process specifically for practitioners of herbal medicine. When choosing a health care practitioner for advice on herbs, look for someone who is a member of a professional organization, such as the American Herbalist Guild.

For information on herbal medicine and referrals to practitioners in your area, contact:

THE AMERICAN HERBALISTS GUILD
1931 Gaddis Rd.
Canton, GA 30115
(770) 751-6021
www.americanherbalistsguild.com

AMERICAN HERB ASSOCIATION
P.O. Box 1673
Nevada City, CA 95959
(530) 265-9552
www.ahaherb.com

THE AMERICAN BOTANICAL COUNCIL
6200 Manor Rd.
Austin, TX 78723
(800) 373-7105; (512) 926-4900
www.herbalgram.org

HERB RESEARCH FOUNDATION
4140 15th Street
Boulder, CO 80304
(303) 449-2265
www.herbs.org

HOMEOPATHY

Homeopathy is practiced by medical doctors (M.D.s), osteopaths (D.O.s), naturopaths (N.D.s), chiropractors (D.C.s), and dentists (D.D.S.s). Some states also allow chiropractors, family nurse practitioners, acupuncturists, and physician assistants to obtain licensures.

For more information on homeopathy or to locate a homeopath in your area, contact:

THE NATIONAL CENTER FOR HOMEOPATHY
801 N. Fairfax Street, Suite 306
Alexandria, VA 22314
(877) 624-0613; (703) 548-7790
www.homeopathic.org

DIET AND NUTRITION

For information on nutrition or to find a qualified nutrition counselor, contact:

THE AMERICAN ASSOCIATION OF NUTRITIONAL
CONSULTANTS
400 Oak Hill Dr.
Winona Lake, IN 46590
(888) 828-2262
www.aanc.net

AMERICAN DIETETIC ASSOCIATION
120 South Riverside Plaza, Suite 2000
Chicago, IL 60606-6995
(800) 877-1600
(800) 366-1655—The Consumer Nutrition Hotline
www.eatright.org

AMERICAN SOCIETY FOR NUTRITIONAL SCIENCES
9650 Rockville Pike, Suite 4500
Bethesda, MD 20814-3990
(301) 634-7050
www.asns.org